PROCESS EQUIPMENT SERIES
VOLUME 1

Solids Separation and Mixing

Edited by
Mahesh V. Bhatia, P.E.
Paul N. Cheremisinoff, P.E.

Contributors to Volume 1

Nicholas P. Cheremisinoff
Paul N. Cheremisinoff
Leonard P. Egee
Ingolf V. Janerus
J. Y. Oldshue
Ronald W. Reid
Robert C. Reisenweber
Alfred Trumpler
W. F. White
Arthur C. Wrotnowski

a TECHNOMIC publication
TECHNOMIC Publishing Co., Inc.
265 Post Road West, Westport, CT. 06880

PROCESS EQUIPMENT SERIES
VOLUME 1

Solids Separation and Mixing

©Technomic Publishing Company, Inc. 1979
265 Post Road West, Westport, CT 06880

a **TECHNOMIC**® publication

Printed in U.S.A.
Library of Congress Card No. 79-63114
ISBN   087762-272-8

FOREWORD

The *Process Equipment Series* is a presentation of exact and useful information relating to equipment and devices used in the chemical and related process industries. The material has been prepared and written by specialists in their respective fields under the editorial direction of the editors.

The equipment, apparatus and devices employed by the process industries are so numerous and varied that the problem was to reduce such information to a moderate and practical size. The editors have given preference to equipment that has general application, and diverse uses in the process industries. The material to be presented has been divided among a series of volumes.

Material presented by authors is primarily written in general technical language and the text has been liberally supplemented with drawings and photographs. The intent has been to provide sufficient theoretical matter to the reader with a satisfactory understanding of the equipment or devices covered. Equipment discussed is of the type that can be purchased by contrast to in-house-designed or constructed apparatus.

The editors express their sincere thanks to the experts who contributed to these collected volumes.

Mahesh V. Bhatia
Paul N. Cheremisinoff

TABLE OF CONTENTS

Page

Chapter 1. Sludge Settling and Thickening 1
Nicholas P. Cheremisinoff, Union Camp Corp. and Paul N. Cheremisinoff, New Jersey Institute of Technology

Chapter 2. Filtration and Dewatering 30
Nicholas P. Cheremisinoff, Union Camp Corp. and Paul N. Cheremisinoff, New Jersey Institute of Technology

Chapter 3. Vacuum Filtration 51
Leonard P. Egee, Komline-Sanderson Engineering Corporation

Chapter 4. Centrifuges 81
W. F. White, Bird Machine Company, Inc.

Chapter 5. Liquid Pressure Filters 105
Alfred Trumpler, Trumpler-Clancy, Inc.

Chapter 6. Particle Classification Using Felt Strainer Bags 175
Arthur C. Wrotnowski, GAF Corporation

Chapter 7. Inclined Plate Settlers 211
Ingolf V. Janerus, Parkson Corp.

Chapter 8. Strainer Selection 228
Robert C. Reisenweber, Zurn Industries, Inc.

Chapter 9. Fluid Mixing 236
J. Y. Oldshue, Mixing Equipment Co., Inc.

Chapter 10. Mixing and Kneading Equipment – Solids and Pastes 267
Ronald W. Reid, Charles Ross & Son Company

Index 292

CHAPTER 1

SLUDGE SETTLING AND THICKENING

NICHOLAS P. CHEREMISINOFF
Union Camp Corp. — R&D Division
Princeton, NJ

PAUL N. CHEREMISINOFF
New Jersey Institute of Technology
Newark, NJ

INTRODUCTION

Water supply quality, waste treatment and retreatment of water supply for reuse are extremely important to industry. The steel, chemicals, petroleum, and pulp and paper industries alone are estimated to utilize roughly 90—95% of all water drawn for industrial purposes. A large portion of this water is extracted and returned to natural surface water courses.

In the past the method of waste disposal consisted of discharging to a water course or lake without treatment. This was particularly true in municipal water treatment. The justification for this practice was that filter backwash waters and settled solids that were returned to natural surface waters added no new impurities, that is, only materials that had been originally presented in the water were returned. However, EPA no longer considers this argument valid since water quality is degraded to the extent that chemicals employed in processing introduce new pollutants. It should be noted that domestic waters are required to be relatively free from materials that may be harmful to the environment and man. They must also meet aesthetic standards with regard to odor, taste and color. On the other hand, requirements for industrial water supplies are markedly different and vary widely, depending on the industry. The chemicals industry, for example, utilizes a major portion of its cooling water in heat exchangers and, as such, influent and effluent waters must meet certain criteria assuring minimum corrosive attack or fouling. In more specialized applications, boiler feedwater must have hardness removed to prevent scaling in boilers.

If the proper treatment program is instituted, contaminants can be removed from a water supply of virtually any initial quality. The drawback is the price in dollars that must be paid. There are a variety of water treatment methods available to the pollution engineer; the specific choice and ultimate design and arrangement of equipment depends on the particular needs and economics.

This section focuses on one particular water treatment method: gravity separation, as it applies to the removal of solid particulates from a slurry or colloidal suspension. Gravity separation techniques are widely employed in both municipal and industrial wastewater treatment. It is by far the most economical method used to achieve particle separation in a colloidal suspension from an energy consumption viewpoint because the separation energy used is gravity. The drawback is the large land area requirements in comparison to requirements for straining of filtration unit operations.

It is the purpose of this section to acquaint the reader with the theory and design criteria behind the gravity separation unit in general, with emphasis on heavy metals treatment. The major process design parameters for selecting and sizing units are discussed and a detailed reference section is included for supplemental reading. Essential to the proper design and operation of gravity separation techniques are the various preconditioning operations required for preparing slurries. Detailed treatment of these processes is given.

SOLIDS DETERMINATION

It is essential that the reader have a clear understanding of the basic terminology and measuring techniques used in gravity settling unit operations. The ultimate design and size of a gravity separation system will depend on the specific application and characteristics of the slurry. These characteristics in turn will establish the criteria for which to apply the theoretical expressions for sizing a unit.

The terms sedimentation and clarification, used interchangeably, refer to the removal of particulate matter, chemical floc and precipitates from suspension through gravity settling. The removal or separation of particulates in suspension occurs either when:

1. the specific gravity of the solid particles is greater than the stream or slurry, resulting in a net downward particle movement; or
2. the specific gravity of the particles is lower than the stream resulting in a net upward particle movement.

Suspended and dissolved solids, both organic and inert, are standard water pollution level tests. Suspended solids are those particulates that can be captured by a fine laboratory filter mat when a sample of the suspension is passed through it. Both suspended and dissolved solids are common parameters employed in defining municipal and industrial waste streams. The operational efficiency of different water treatment systems is defined in terms of the solids removal. The operational efficiency of a gravity settler is defined in terms of the suspended solids removal in a settling basin: Suspended solids are of prime interest to gravity separation because they are the only particulates that can be removed by gravity separation, often without any pretreatment.

Total solids or total residue is the sum of the dissolved and suspended solids. This quantity is normally measured by allowing a known aliquot of slurry sample to evaporate in a porcelain dish and then drying it in an oven. The remaining residue is the total solids. Normally, evaporation is accomplished by indirect steam heating; then the dish is transferred to an oven and dried at uniform temperature (the standard being 103°C) to a constant weight. The weight of total residue equals the difference between the cooled weight of the sample dish and the original weight of the empty dish. The concentration of the total residue is computed from the following formula:

$$C_T = \frac{W \times 1000}{V} \qquad (1)$$

where C_T = concentration of the total residue, mg/l
W = total residue weight, mg
V = volume of the slurry sample tested, ml

Some confusion does exist among testing standards and the reader should be aware of these. The terms described apply strictly to wastewater analysis. However, when dealing with water analysis in general, those particles which pass through a filter pad are referred to as nonfiltrable residue and those that are retained on the pad are called filtrable.

Suspended solids are normally tested for with a Gooch crucible, which is a small porcelain filtering dish with tiny holes in the bottom and an asbestos fiber mat. A measured portion of the slurry is transferred to a funnel and drawn through the filter pad by applying a vacuum to a suction flask. The disc is removed from the funnel after filtering, dried and weighed to determine the new weight due to the residue retained. The total suspended solids concentration can be computed from a formula with the same general format as Equation 1. The dissolved solids can be determined from this measurement by subtracting the suspended solids from the total residue on evaporation.

Often it is necessary to remove various negative and/or positive ions of the dissolved solids in the slurry. To accomplish this by gravity settling, chemical additives are added to react with the dissolved solids in the slurry to cause precipitation. The precipitate consists of a compound having low solubility in the liquid. Hence, dissolved solids are transformed to suspended solids for removal by gravity separation.

Further analysis can be performed on the slurry sample such as the determination of the amount of inerts and volatiles in the solids. Volatile samples can be determined by igniting the total collected residue after evaporation. This is normally done in an electric muffle furnace at 550°C. Sludge samples are incinerated for a period of one hour whereas, water samples require 20 minutes. The volatiles are reported as the weight loss of the sample per unit volume of sample tested (i.e., mg/l). The remaining residue is referred to as the fixed solids. Volatile solids determination is usually considered as being a measure of the inorganics. It is important to note that this interpretation is not entirely accurate since many organic compounds leave an ash when incinerated. In addition, many inorganic salts volatilize during burning.

A measure of the settleable solids can also be made on slurry samples. These are determined by filling a one-liter conical-shaped flask (called an Imhoff cone) and allowing a certain time span (usually one hour) for sedimentation. The tip at the cone's bottom is marked for volume indications so that the quantity of settled matter can be recorded directly in ml/l of sample. The settleable matter may also be expressed on a weight basis. This is done by putting a sample of the wastewater into a one-liter graduated cylinder and allowing it to stand quiescent for a period of one hour. Solids analysis must be run on the initial well-mixed sample and on the suspended particulates remaining at the midpoint of the cylinder after settling. By use of the following expression, the mg/l of settleable solids can be computed:

$$C_S = C_{SS} - C_{NS} \qquad (2)$$

where C_S, C_{SS} and C_{NS} are the concentrations, in mg/l, of the settleable suspended solids and nonsettleable solids, respectively.

Solids concentrations reported in mg/l are generally low concentrations (in terms of ppm in the range of 0–1,000). Intermediate concentrations are reported in terms of ppm and high concentrations in percent solids. The definition of ppm and percent solids is given in Equations 3 and 4, respectively.

$$\text{ppm} = \frac{\text{weight of solids}}{\text{weight of slurry}} \times 10^6 \qquad (3)$$

$$\text{Percent Solids} = \frac{\text{weight of solids}}{\text{weight of slurry}} \times 100 \qquad (4)$$

Note that ppm can be directly converted to percent solids by dividing ppm by 10^4. When the solids concentration is low, the specific gravity of the slurry is approximately one, and, hence, mg/l can be assumed equivalent to ppm.

BATCH SETTLING

To illustrate what occurs during a settling operation, consider a slurry consisting of finely divided particles of uniform density and spherical in shape, contained in a vertical transparent cylinder. If the suspension is dilute, particles can be observed to settle down through the liquid at a rate dependent upon the particle size, the viscosity of the liquid and the relative density of particle and liquid. Eventually, the liquid will become clear and a pile will collect at the bottom. Careful observation during the settling time will reveal no clear line of demarcation between clear fluid and the settling slurry.

More concentrated slurries, hoever, will behave differently. Solids tend to settle at a slower rate, primarily because of interference between particles. In addition, particles tend to be more uniform in size and settle together. One can then observe a reasonably sharp boundary between the supernatant liquor in the upper region of the cylinder and the settling region in the lower portion.

The careful observer will notice that four distinct zones exist at any one point in time during settling. Figure 1.1 illustrates what happens at one point in time during the experiment by a plot of the particle density (number of particles per unit time) vs the height of the cylinder. Zone 1 indicates the clear supernatent liquid. In Zone 2, the solids concentration is fairly uniform. As shown by the graph, the particle density is at a uniform value, n_o. Zone 3 is a transition zone between 2 and 4 and is not always sharply defined. Zone 4 is usually referred to as the compression zone. In this region particles accumulate on top of each other and the liquid is forced out from between the layers of particles due to their weight. As shown by the curve, for compressible sludges, this results in increasing solids concentration with depth (note that average solids density was greater than the liquid, the opposite would be true). Note that the heights on the ordinate will change with time. At some time later, h_3 will be higher up the axis and eventually Zone 2 will disappear.

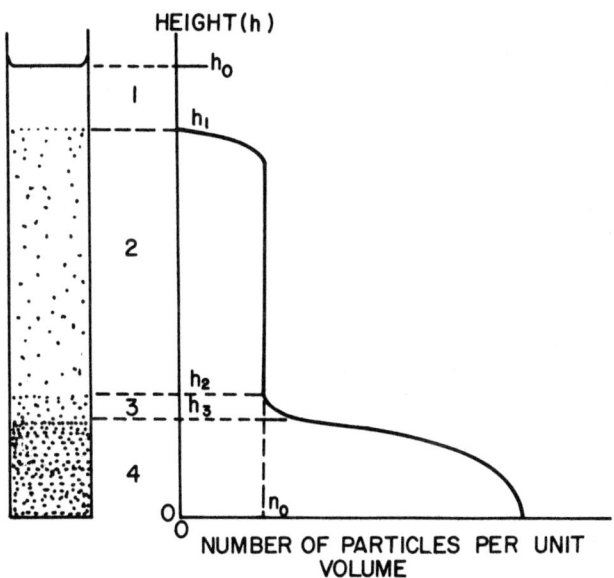

Figure 1.1 Four distinct zones can be observed in batch settling.

The rate at which settling occurs can be recorded by a plot of the height of the boundary between Zones 1 and 2 vs time or by a plot of the upper height of Zone 4 vs time. Such a plot is shown in Figure 1.2. The top of Zone 2 descends the

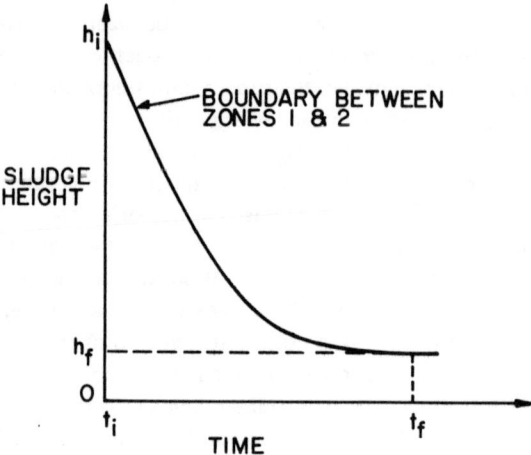

Figure 1.2. Rate of solids settling, where h_i is the initial height of the top of zone 2 at time t_i, when the experiment initiates, and h_f is the final height at time t_f, when settling is completed.

cylinder at a steady rate (*i.e.*, solids settle at a constant rate) until the zone has almost disappeared and all the suspended solids are in the compression zone. The rate at which the height of Zone 4 rises, or rather the rate of settling in the compression zone is not constant and is relatively slow. In other words, two settling rate periods are often observed. It should be noted that the shape of the curve in Figure 1.2 does not always take this general form. Settling curves depend on the physical and chemical characteristics of both the solids and liquid, particle sizes, the concentration of the slurry, and also the initial height when observations are initiated.

MIXING AND FLOCCULATION

Chemicals are used in the early stages of industrial and municipal waste treatment for a number of reasons:

1. phosphorous removal in primary clarification,
2. to achieve greater removals of Biological Oxygen Demand (BOD),
3. to achieve proper preconditioning of wastes for filtration, carbon absorption or reverse osmosis,
4. for pH control or
5. to effect primary removal of suspended solids in the primary clarification step.

For surface water treatment, the basic flowchart of a treatment system would consist of clarification by coagulation, sedimentation and filtration. Figure 1.3 shows the primary process. Chlorination is often done in the first step to disinfect

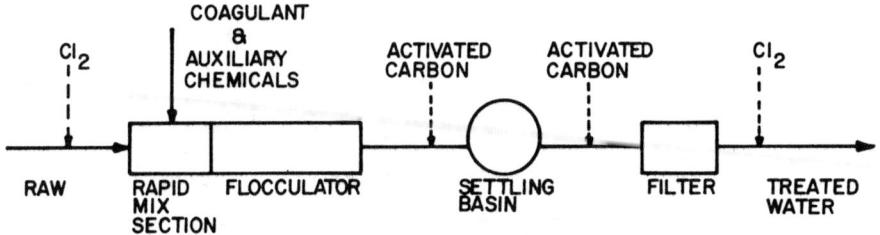

Figure 1.3. Basic flowchart for surface water treatment system.

the raw water. It may also be done to the effluent stream to establish a chlorine residual. Activated carbon is used to remove taste- and odor-producing matter. It should be noted that Figure 1.2 is only a very basic scheme. Many surface water treatment systems have two or more stages of chemical coagulation and sedimentation. Such arrangements are operated in series, or can be split in a parallel arrangement with softening in one stage and coagulation in another.

The major sources of waste from the process are sludge from the settling basin and washwater from backwashing filters. These wastes can vary considerably in composition and concentration on a daily basis. In addition, although these wastes are produced on a continuous basis, they may be discharged intermittently.

Consider again the example of sludge settling in a transparent cylinder. If the finely divided particles have similar electric charge, they tend to repel each other and the suspension remains dispersed. The addition of a flocculating agent, which itself is an electrolyte, will tend to neutralize the particles. This causes particles to form flocs or aggregates.

The finely divided particles in a slurry are referred to as colloids. Particle sizes can range anywhere from 1–500 mμ. Under such conditions particles will not settle out without the use of some coagulant or flocculating agent because the individual particles have such a large surface area in comparison to their weight that gravity has no effect on them. There are two classifications of colloids: (1) hydrophobic or water hating, and (2) hydrophilic or water loving.

The latter is much more stable because these colloids have an affinity for water molecules (examples are soaps and detergents). Hydrophobic matter, however, derive their suspension stability from their own electrical charge. Figure 1.4 illustrates the attractive and repulsive forces that are at play with hydrophobic particles. Electrostatic repulsion forces develop because positive ions in the suspension adsorb onto the particles' surfaces. The magnitude of these repulsive forces is called the zeta potential. Because of the force of the zeta potential and the opposing force of ions of different charge in the liquid, the force of gravity no longer becomes the dominant parameter affecting stability.

It should be noted that other forces are at play in Figure 1.4. Van der Waals forces is the natural attraction that exists between two bodies. Brownian movement

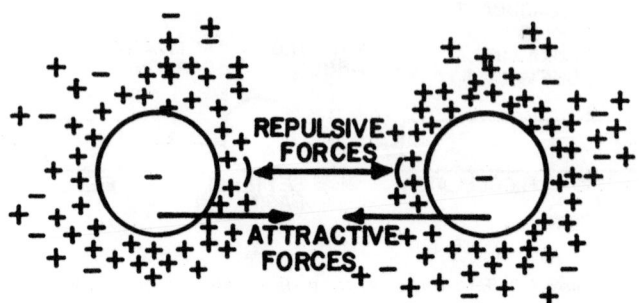

Figure 1.4. Hydrophobic particles in a stable suspension.

is the random movement of particles caused by interaction or bombardment of the water molecules. Although these forces tend to enhance attraction, colloidal suspensions will still remain stable in solution because the repulsive forces exceed those of attraction so that particles rarely come into contact.

To achieve settling of a colloidal suspension, the zeta potential and/or the opposite ionic charges in the slurry must be reduced. This is done by destabilizing the hydrophobic colloids through the use of high positively charged coagulating agents (most colloidals are negatively charged). The most common chemical coagulants are salts of aluminum and iron. In solution these salts produce highly charged hydrolyzed metal ions (trivalent ions of aluminum (Al^{+++}) and iron (Fe^{+++}) for the example given; usually the salts are sodium aluminate ($NaAlO_3$), alum ($Al_2(SO_4)_3$), ferric sulfate ($Fe_2(SO_4)_3$), or ferric chloride ($FeCl_3$). These metal ions reduce the forces of repulsion around the individual particles by compressing the diffuse double layer. Once these forces have been reduced, slowing mixing enhances particle contact and the attractive forces cause particles to adhere to each other. Figure 1.5 illustrates what takes place.

Organic chemicals known as polyelectrolytes are also used for coagulation. Polyelectrolytes are high molecular weight, water-soluble polymers and, when in solution, they undergo electrolyte dissociation resulting in highly charged ions with long chains causing colloid particles to adhere to their surface. Figure 1.6 illustrates what happens when a polyelectrolyte is added to a colloidal suspension. Polyelectrolytes are most often used as a coagulant aid for the inorganic chemicals; however, there are many applications where they are used alone.

Slow stirring of a colloidal suspension with coagulants, can greatly alter the rate of settling. Stirring alters the floc structure such that the solids concentration in Zone 2 of our transparent cylinder would no longer be uniform. Under such conditions Zone 4 would be observed to have a much sharper line of demarkation.

The selection of specific coagulants depends on the nature and characteristics of the colloids; the temperature and pH of the slurry; the chemical composition of the water or liquid; and economic consideration. The rate of solids removal will depend

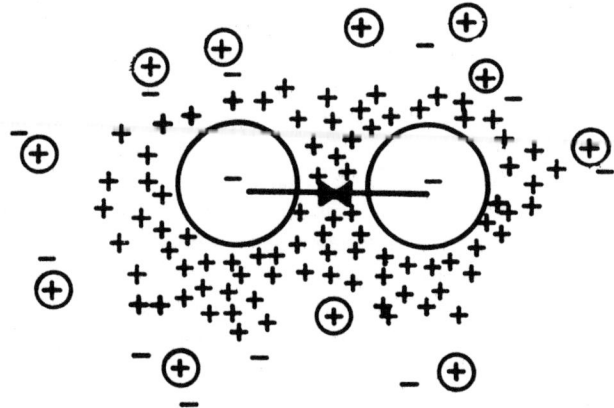

Figure 1.5 Destabilization of repulsive forces by the addition of positively charged metal ions.

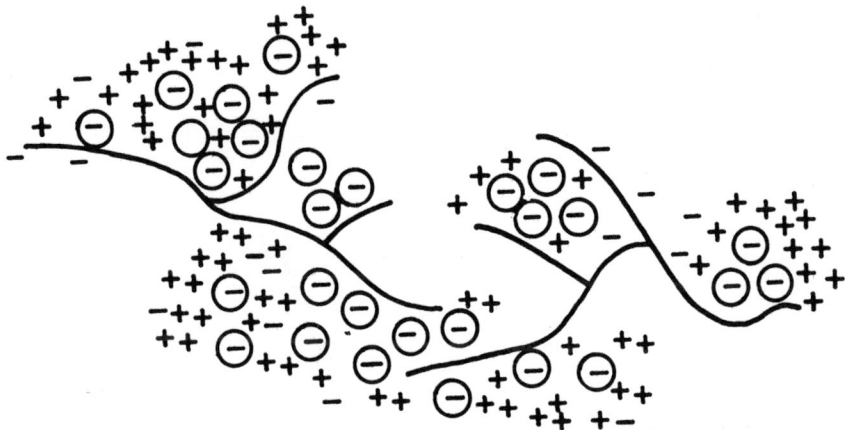

Figure 1.6 Long-chain polyelectrolytes cause agglomeration of colloid particles.

on all these factors plus the type and dosage of coagulants and aids used in addition to the time and degree of mixing provided to achieve good chemical dispersion and floc formation. It is important to note that the term coagulation refers to the destabilization of a colloidal dispersion by suppressing the double-layer charge, whereas flocculation means an aggregation of particles. However, our discussion is concerned primarily with the physical processes of chemical treatment rather than the chemistry, and, as such, no distinction is made between the two terms.

Agitation or mixing enhances coagulation and settling by dissolving and dispersing chemical additives throughout the slurry. This is generally a rapid mix operation. Slow or mild mixing enhances flocculation by allowing well-developed flocs to form and settle. For the purpose of this discussion, coagulation will refer to the entire process of mixing and flocculation. As shown in Figure 1.3, the entire process may be housed in one unit. These units are designed as either constant-stirred tank reactors (CSTR) or plug flow reactors.

In an ideal CSTR, the influent is dispersed immediately throughout the basin volume. The concentration of the reactant in the effluent stream is the same as the concentration in the vessel (refer to Figure 1.7). in water treatment applications, this type of reactor is referred to as a rapid mix tank (or flash mix tank). It does not necessarily have to be a separate unit, but rather a section of the mixing and reaction section in flocculator-clarifiers. The detention time in this portion of the flocculator-clarifier is given by the following expression:

$$\tau = \frac{1}{k}\left(\frac{C_o}{C_t} - 1\right) \quad (5)$$

where τ = detention time of the basin (or portion of the basin occupied by the mixing section, s
k = rate constant, s^{-1}
C_o,
C_t = influent and effluent reactant concentration, ppm or mg/l, respectively.

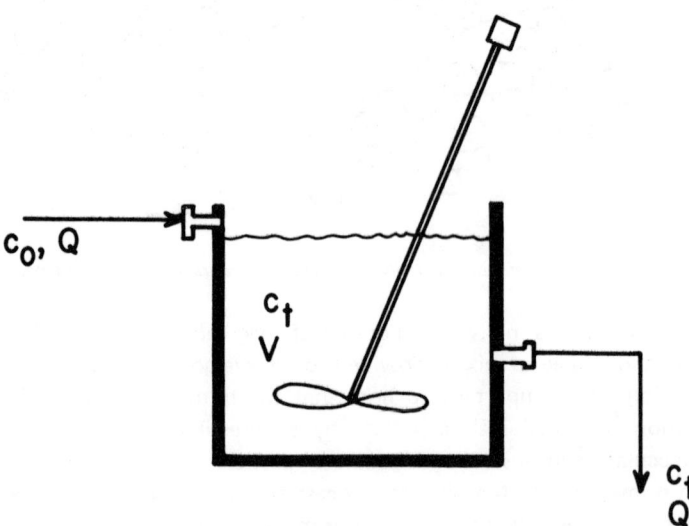

Figure 1.7 Ideal constant-stirred tank reactor under steady-state conditions.

Detention time is defined by the following:

$$\tau = V/Q \qquad (6)$$

where V = volume of the basin, m^3
       Q = volumetric flowrate of the slurry, m^3/s

An ideal plug flow reactor is shown in Figure 1.8. As shown, the concentration of reactant decreases as a function of reactor length. An ideal reactor of this type can be thought of as a long tube in which an imaginary plug of slurry is moving through. Plug flow conditions are very difficult to achieve in practice because of turbulence generated by frictional resistance from the tube wall. The use of baffles in flocculation tanks can reduce eddying and near plug flow can be approached. For first-order kinetics, the detention time in the vessel is given by:

$$\tau = \frac{1}{k}(\ln(C_o C_f)) \qquad (7)$$

where k and C_o are the same as in Equation 5 and C_f is the effluent reactant concentration.

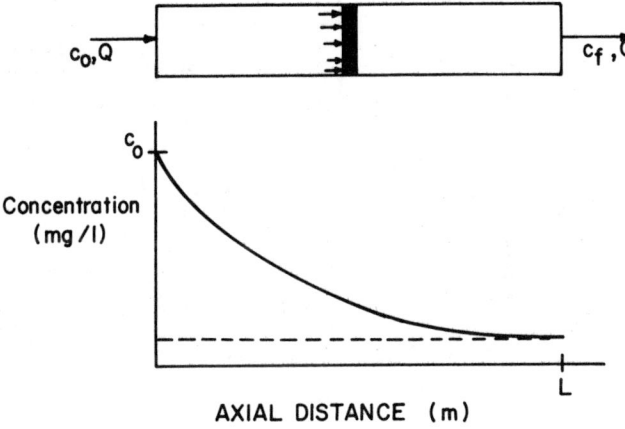

Figure 1.8 An ideal plug flow reactor under steady-state conditions.

The detention time is defined the same as in Equation 6; however, because of the geometry of this reactor, the expression reduces to the following:

$$\tau = \frac{V}{Q} = \frac{AL}{vA} = \frac{L}{v} \qquad (8)$$

where A = cross-sectional area of the reactor m^2
       L = length of the reactor, m
       v = linear velocity of flow through the reactor, m/s

The reader should refer to Levenspiel (1) for a more detailed treatment of reactor design and kinetics.

Horizontal and vertical redwood paddle wheels are conventional designs used in round or rectangular flocculation basins (CSTR). Other arrangements include:
1. vertical turbine flocculators (adaption from center well reactor clarifier units);
2. horizontal turbine flocculation;
3. horizontal paddle oscillating flocculators; and
4. air flocculation.

Cheremisinoff and Young (2) give a detailed discussion of these and other deisgns.

A typical vertical turbine flocculator in a high-rate reactor clarifier is shown in Figure 1.9. A turbine type flocculator is shown in Figure 1.10. These units are substantially smaller in diameter than paddle wheel flocculators. It operates based on the pumpage of flow radially outward and has a recirculation of flow up to five times within the detention time of the flocculation zone. These units have been widely used for flocculating fine particulates (colloidal solids at high zeta potential).

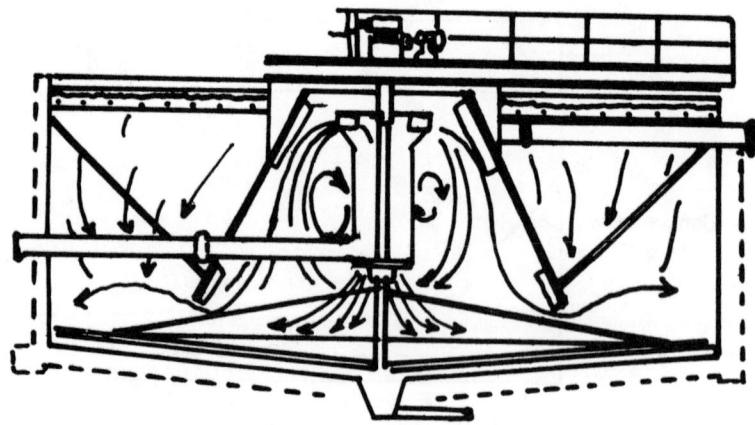

Figure 1.9. Vertical turbine flocculator in a high-rate reactor clarifier.

There are basically two types of mixing operations: rapid mix and slow mix. Rapid mixing is done to prepare a homogeneous blend of coagulants and slurry. In slow mix operations, sufficient particle movement is provided to ensure a high probability of particle collision. Collisions enhance the formation of agglomerates with higher specific gravities and greater settling velocities. Rapid mix designs have

Figure 1.10. Vertical turbine flocculator.

short detention times (on the order of 1–10 minutes) and high agitation horsepower unit volume. Slow mix operations have much longer detention times (10–30 minutes) and significantly lower mixing horsepower per unit volume. A detailed treatment of mixing theory is given by Camp. (3)

It is worthwhile at this point to end our discussion by outlining some general recommendations for flocculation basin design. An illustrative example follows outlining the use of the major equations presented here. In general, flocculation basins should be designed and sized with the following in mind (4):

1. Minimum slurry flow through velocities should not be < 0.15 nor > 0.46 m/min.
2. Detention times should not be < 30 minutes for floc formation
3. The inlet and discharge zones should be designed in such a way so they do not produce unwanted turbulence that can break up flocs
4. Agitators or paddles should be driven with variable-speed drives. The peripheral speed of paddles should operate in the range of 0.15–0.76 m/sec.
5. Flocculation and sedimentation units should be mounted as close as possible to each other
6. Flocculated slurry velocities through transmission lines to settling basins should be in the range of 0.15–0.46 m/sec.
7. Allowances and precautions should be taken to reduce turbulence-generated expansions, contractions, bends, etc. in transmission lines.

Illustrative Example.

Calculate the detention time in a CSTR for a chemical coagulation reaction with an initial reactant concentration of 380 mg/l and a 95% reduction. Laboratory bench-scale studies determined the reaction to follow first-order kinetics with a rate constant of 220 day^{-1}.
Solution:

For a CSTR use Equation 5) —
$C_o = 380$ mg/l, $C = 380 - 0.95 (380) = 19.0$ mg/l

$$\tau = \frac{\text{day}}{220} \times \frac{1440 \text{ min}}{\text{day}} \times \left(\frac{380 \text{ mg/l}}{19 \text{ mg/l}} - 1 \right)$$

$\tau = 124$ min

SEDIMENTATION

Sedimentation was previously defined as the removal of suspended solids by gravity settling. Separation of rising solids in slurry is referred to as flotation. Those solids that will readily settle or have reasonable settling velocities without chemical pretreatment are referred to as settleable solids. Solids are considered settleable if individual particle settling velocities are greater than or equal to 0.037 ft/min.

There are three types of settling processes:

1. discrete settling,
2. flocculant settling, and
3. hindered settling.

In discrete settling, the slurry is characterized by low suspended solids concentration. Hence, the settling velocity of individual particles is not greatly influenced by interaction or contact with neighboring particles. This means then that the settling velocity of a particular particle is constant through the distance it falls. The settling velocity is solely a function of its shape, mean diameter, specific gravity and the properties of the slurry (viscosity and specific gravity). For discrete particle settling, the ideal settling basin is shown in Figure 1.11. For simplicity the chamber is considered rectangular having length, L, width, W, and height, h. From the illustration the following expressions are readily derived:

$$V = \frac{Q}{A} = \frac{Q}{hW} \qquad (9a)$$

$$\frac{h}{V_p} = \frac{L}{Q/hW} \qquad (9b)$$

$$V_p = \frac{Q}{LW} \qquad (9c)$$

where   A = cross-sectional area of the chamber in the direction of slurry flow
        Q = volumetric flowrate of the slurry
        v_p = resultant particle velocity ($v_x + v_s = v_p$)
        V = linear velocity of the slurry through the chamber.

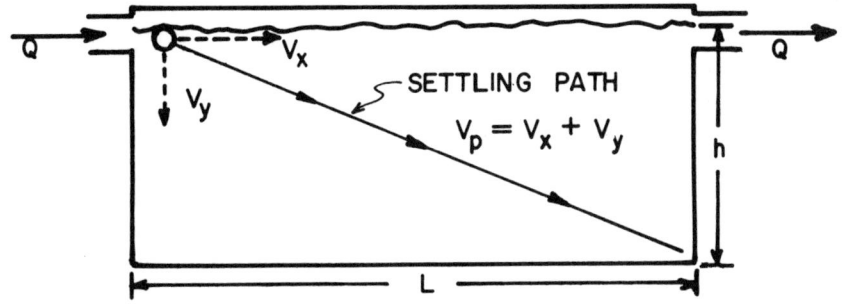

Figure 1.11. Ideal settling chamber.

Note that v_x is the tranverse velocity component of the particle and v_y is its settling velocity. A detailed treatment of discrete particle settling is given by Cheremisinoff. (5) From Equations 9a–9c we obtain:

$$\frac{h}{v_p} = \frac{L}{V} \tag{10}$$

The underlying assumption for our idealized settling chamber is that suspended solids in the influent stream are homogeneously mixed (*i.e.,* the inlet stream has an even particle distribution), although in practice particulates vary in size and settling velocities.

Settling chambers are designed to operate in a specified certain range of particle settling velocities. That is, a criterion for sizing is to design a unit for removing all particles with settling velocities greater than and equal to a specified value. For the purposes of our discussion assume we wish to remove all suspended particulates having one settling velocity (v_p^1). The critical influent position for a particulate with v_p^1 is at the slurry level at the chamber inlet. The particle will have a settling path that results from its transverse velocity and settling velocity, as shown in Figure 1.11. For the particle to come to rest at the intersection of the floor and the far wall of the chamber, the time it takes to travel the distance, L, must equal the time to fall through height, h. The time to travel over L depends on the volumetric flowrate (referred to as the hydraulic flowrate, Q) and the time required to travel through h depends on v_s.

Equations 9 and 10 reveal that if Q and v_p are known (where $v_p = v_p^1$, the settling velocity required to remove our particle), then the basin can be sized in terms of its surface area.

The particle velocity, v_p, is referred to as the "surface loading" or "overflow rate." The units of v_p are obviously ft/min; however, common practice is to use gal/day/ft² (gpd/ft²) or gal/min/ft² (gom/ft²). (Note ft/min × 7.5 gal/ft³ × 1440 min/day gpd/ft²). It should be noted that the height of the chamber, h, is not pertinent to the design. The removal of particles with a certain settling velocity depends on LW.

It is obvious that all particles with settling velocities greater than v_p^1 will also be removed; however, these particles will settle onto the chamber floor at various distances from the effluent wall. Less obvious is that a certain portion of the particles with settling velocities less than v_p^1 also settle out. The fraction of those particles with velocities less than v_p^1 that will be removed depends on the depth at which they enter at the basin influent end of the homogeneous slurry. The number of particles is directly proportional to its settling velocity and inversely proportional to settling velocity for which the basin is designed. Mathematically this is expressed by:

$$X = \frac{C_R}{C_M} = \frac{v_M}{v_D} \tag{11}$$

where X = fraction of particles having concentration C_M removed
C_R = concentration of particulates removed
v_M = measured or estimated settling velocity of particles with concentration C_M
v_D = settling velocity for which the basin was designed

The following example illustrates the use of the expressions presented thus far.

Illustrative Example 2

A homogeneous slurry is assumed to undergo discrete settling without any chemical pretreatment. Samples of the slurry were tested for particle size distribution, suspended solids concentration and settling velocities. The data are given in the Table 1.1.

Table 1.1. Laboratory Analysis of Slurry

Settling Velocity (fpm)	Particle Size (μ)	Fraction of Total Suspended Solids at Indicated Velocity
0.2	50–60	0.10
0.3	80–90	0.11
0.6	100	0.35
0.9	120	0.28
1.4	150	0.16
		1.00

1. Size the settling basin for an overflow rate of 6480 gpd/ft² and a hydraulic loading of 3×10^6 gpd.
2. Compute the detention time in the basin.
3. Based on the above overflow rate, determine how much settleable solids is removed in ppm if the influent concentration is 700 ppm.

Solution:

(a) Using Equation 9c:

$$A = LW = Q/v_p$$

$$= \frac{3 \times 10^6 \text{ gpd}}{6480 \text{ gpd/ft}^2} = \underline{463.0 \text{ ft}^2}$$

This is the cross-sectional area of the basin required for the specified overflow rate. The actual dimensions will depend on space limitations. For the purposes of this example, we select the following dimensions:

$$W = 12 \text{ ft}, L = 38.6 \text{ ft}, h = 8 \text{ ft}$$

(b) Detention time is defined as by Equation 6:

$$V = L \times W \times h = 3704 \text{ ft}^3$$

$$T = \frac{V}{Q} = \frac{3704 \text{ ft}^3}{3 \times 10^6 \text{ gpd}} \times 7.5 \frac{G}{\text{ft}^3} \times 1440 \frac{\text{min}}{\text{day}} = \underline{13 \text{ min}}$$

(c) It is desired to remove particles based on the specified overflow rate:

$$\frac{6480 \text{ gpd/ft}^2}{7.5 \text{ g/ft}^3 \times 1440 \text{ min/day}} = 0.60 \text{ fpm}$$

Hence, all particle concentrations having settling velocities greater than and equal to 0.60 fpm will be removed completely. The concentration of suspended solids with settling velocities less than 0.6 fpm can be computed from Equation 11. Calculations are tabulated below.

Sample Calculation — At 0.6 fpm; $C_M = (0.35) \, 700 \text{ ppm} = 245 \text{ ppm}$ the solids concentration at the influent; 100 of this is removed.

At 0.03 fpm; $C_M = 0.11 \, (700 \text{ ppm}) = 77.0 \text{ ppm}$ the solids concentration at the influent; From Equation (11)

$$C_R = C_M \frac{v_M}{v_D}$$

$$= 77 \text{ ppm} \times \frac{0.3 \text{ fpm}}{0.6 \text{ fpm}}$$

$C_R = 38.5$ ppm of the solids with this settling velocity are removed.

Summary of Calculations:

Fraction of Total SS at Indicated Velocity	Settling Velocity (fpm)	Actual Inlet Concentration (ppm)	Concentration of SS Removed (ppm)
0.10	0.2	70.0	23.3
0.11	0.3	77.0	38.5
0.35	0.6	245.0	245.0
0.28	0.9	196.0	196.0
0.16	1.4	112.0	112.0
		700.	614.8

Hence, the fraction of SS (suspended solids) is $614.8/700 = \underline{0.88}$

It is necessary to have reliable estimates of particle settling velocities for various concentrations or solids sizes in slurries to be treated. Settling velocities can be directly measured by tall cylinder laboratory tests. This measurement technique employs a tall graduated transparent cylinder ranging anywhere from 4–8 ft in height and is normally applied to discrete particle settling tests with velocities less than 0.91 cy/sec. Sampling points are located at various distances from the top of the cylinder to allow measurements of settling and sludge concentrations.

Briefly, the method involves the following steps:

1. The slurry is analyzed for suspended solids concentration and a known amount is poured into the cylinder. The sample is thoroughly mixed to assure homogeneity.
2. A small amount is withdrawn at each sampling port. The height and time for each withdrawal is recorded.
3. Each sample is analyzed for ss concentration. These concentrations represent the amount of ss remaining in the slurry sample after the recorded time (since the height at which each sample was analyzed is known, this also represents the amount of ss remaining after a known settling velocity. This information is expressed in the form of a plot as ss concentration vs settling velocity and enables the design engineer to determine the concentration that would remain in the settling chamber effluent for a specified settling velocity. As outlined in Illustrative Example 2, the overflow rate specified from the chosen settling velocity is divided into the design hydraulic flow-rate to determine the settling chamber surface area. Usually the settling velocity is multiplied by a safety factor (1.3–1.6) to establish the overflow rate.
4. A portion of the sludge at the bottom sampling port is withdrawn for solids analysis. This provides information solid balances that are useful in selecting and sizing under-flow pumps, lagoons, etc.

A second approach to obtaining settling velocity information is to approximate values from Stokes law. This is only used as a rough estimate; however, it gives reasonaably good values for particle sizes in the range of 65–150 mesh (in terms of settling velocities, in the range of 0.91–1.90 cm/sec). This expression is derived by a force balance on a spherical particle. For a discussion of its theoretical development the reader should refer to the literature (4):

$$v_y = \frac{S(G_S - G_L)D^2}{\mu} \quad (12)$$

where C is a shape factor (for a spherical particle, C=1; for the shape C<1); G_S and G_L are the specific gravity of the solid particle and liquid, respectively; D is the particle mean diameter; and μ is the viscosity of the liquid. Examples of the use of Stokes law in estimating settling velocities can be found in the literature (5,6).

Type II Settling

In type II or flocculant settling, conglomerated particle characteristics change as a function of time in the settling chamber. This means then that the velocities of particles are not constant but rather increase with time and distance through the chamber. If we view a flocculant as a spherical particle, as in our example for discrete settling, then the settling path becomes distorted and resembles Figure 1.12. Since velocity is a variable for this type of settling, the quantity of ss remaining in the exit stream will depend not only on the surface area of the chamber, but the height, h, as well.

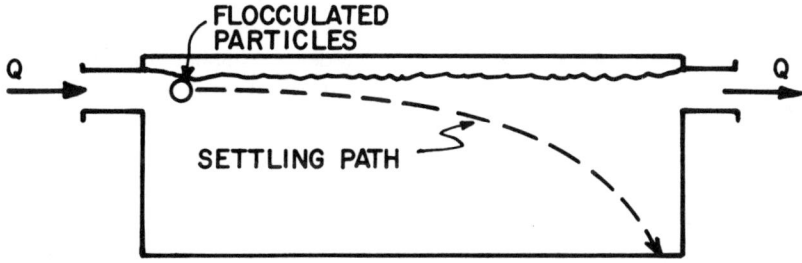

Figure 1.12 Settling path of flocculated particles in an idealized settling basin.

Careful distinction should be made between floc particles and those having varying settling velocities. Floc particles generated by either chemical precipitation or coagulation will have properties similar to discrete settling particles because although settling velocities vary during the slow mix operation, the floc is reasonably stabilized by the time it reaches the influent of the settling chamber. Hence, floc particle settling velocities become constant. To distinguish between the two types of settling, tall cylinder settling tests should be conducted. Briefly, tests should be conducted by sampling from a series of levels in the cylinder at each time interval. The solids concentration in each portion collected should be analyzed with the surface loading parameter and the depth variable. This information can be applied to estimating design parameters for a specified minimum concentration of remaining solids in the effluent stream.

Type III Settling

Hindered settling is characterized by relatively high solids concentrations in the liquid such that there is complete particle contact in the slurry. Settling takes place in concentrated layers rather than individual particles. This is the most complicated type of settling from a modeling point of view in that settling is a function of the separate concentration layers. A crude indication of hindered settling is to put a sample of the slurry in a clear beaker or cylinder. If it is a hindered settling slurry, distinct interfaces can be observed between various layers of solids concentration.

An additional parameter that will have to be tested for is the possibility of limiting solids concentration layers. In our cylinder, as the solids settle they pass through a range of concentrations (from the initial concentration of maximum values). The same will occur in a settling chamber. The problem may arise that any of these concentration layers can be capacity limited. In other words, if the rate of solids in the influent layer is greater than the downward rate of that particular layer, the gravity settling unit will not function in the range for which it was designed. For hindered settling slurries, both the overflow rate and the solids loading must be the basis for the design.

There are a number of methods employed in obtaining design parameters for this type of settling. The most widely used laboratory technique employs 1 1. graduate for analyzing solids concentration layers during quiescence. A slow rpm mechanical stirrer is stationed at the bottom of the cylinder to simulate the influence of mechanical sludge collecting mechanisms (called rakes) employed during actual chamber operations. The method involves the following steps: (1) The slurry of known solids concentration is poured into the cylinder and the sitrrer rotated; (2) the level of the solid interface is noted as a function of time; (3) from the interface height-time data a curve can be constructed as shown in Figure 1.13. As shown in the plot, three distinct and constant ss concentration zones will develop; and (4) samples are periodically withdrawn from the compression zone to check for the maximum concentrations achievable in the underflow sludge (5).

The plus shown in Figure 1.13 can be applied to evaluating the design parameters of surface loading and critical solid loading. Critical solid loading is normally expressed in units of lb of solids charged to the settling chamber/day/ft^2. The surface loading parameter is characterized by Zone 1 — the constant concentration zone. The slope of this portion of the curve (which is approximately a straight line and represents the solid-liquid interfacial position of the initial solids concentration), represents the settling velocity of the initial solid concentration. By using an appropriate safety factor this can be converted into the surface loading.

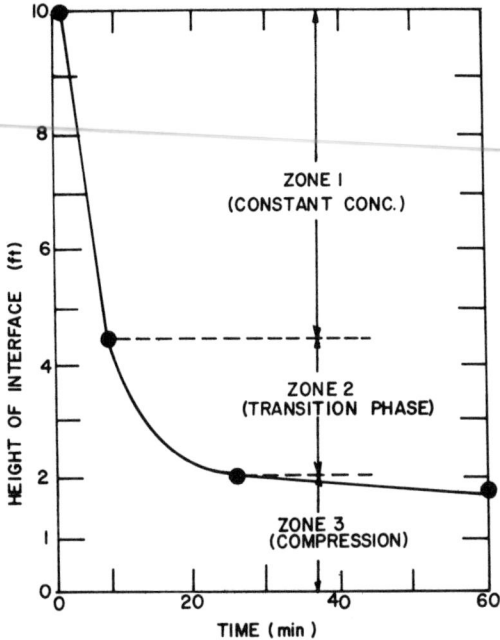

Figure 1.13 Hindered settling curve showing three zones.

Often hindered settling curves display characteristics that indicate flocculant properties of solids in slurries. Most biological slurries for example have natural flocculant properties and, as such, this is a characteristic of the slurry. This is observed on the plot by a curved portion over a short period near the beginning of the constant concentration zone. For design estimates, the short curve section is usually compensated for by adjusting the slope of the linear portion of the curve in this zone.

FLOTATION

Normal practice in designing flotation units for ss removal is to apply gravity separation theory to obtaining rise velocities of particles. Precautions must be taken and adequate safety factors used in applying design parameters.

A more common practice is to induce flotation by introducing fine gas bubbles into the raw water (known as Dissolved Air Flotation). Particles with specific garvities near 1 adhere to gas bubbles and become buoyant. In municipal wastewater treatment, gas bubbles are relatively fine — in the range of 0.01–0.1 mm.

Vacuum flotation is another technique used for introducing gas bubbles. Both methods supersaturate the raw water with air at a certain pressure. Flotation has several advantages over sedimentation. The main advantage being that it makes use of higher surface loading and, as such, requires smaller tank sizes. Recall that the large surface area requirements for sedimentation basins is its major disadvantage. Second, flotation is a much more effective technique in removing solids which are otherwise difficult or too costly to settle with coagulants. In addition, these designs have the ability to handle peak seasonal loads and/or storm flows. They are often used to intermittently increase the capacity of settling basins.

Dissolved air flotation techniques have been proposed for use in grit and scum seapration in single treatment units. Design parameters and performance data for such applications are scarce. A detailed treatment of this subject is given in the literature (7,8).

Dissolved air flotation systems generally use rectangular vessels equipped with separate chain-and-flight scum and sludge collectors. As with settling tanks, designs may also be circular. These units are primarily employed as thickeners for activated sludge. Fouling of pressuring and pressure-regulating equipment is a major problem. To minimize scaling and fouling, a stream of recycled effluent is pressurized. The stream is then blended with the influent on pressure release.

The mining industry has had years of experience using dissolved air flotation systems for concentrating mineral ores. The paper industry also employs such units for treating white water for fiber and water recovery. Other applications include treatment of oily wastes from refineries, steel mills, railroad terminals and petrochemical plants. In these types of slurries the oil tends to coat solid particulates and retards settling. Air bubbles tend to enhance agglomeration and induce flotation.

There are three types of flotation units (Figure 1.14). Partial aeration systems pressurize only a portion of the waste stream (approximately one-third). When the solids load is light, this greatly reduces pumping costs. For direct aeration the entire influent stream is pressurized and aerated with air. The material to be removed must be capable of withstanding shearing forces in pressure pumps and pressure release valves. Effluent recirculation, already described, is the most commonly employed air flotation system. It is normally employed with coagulants or flocculants, particularly for those materials that cannot withstand high shearing forces.

As in sedimentation, flotation aids are used to enhance separation. Flotation aids are employed for several reasons. The addition of flocculants increases attraction between particulates and air bubbles. This is accomplished by either increasing particle size or by the reduction of solids charges. In many metallurgical flotation systems, chemical aids are added to accelerate separation rate.

Tall cylinder tests should be conducted to evaluate various flotation aids effects and rising velocities. It is important to apply marginal safety factors to these data in scaling up designs, especially since these tests do not entirely recreate the degree of turbulence and agitation that occurs in transmission lines or in flotation units. The particular chemical additives selected must produce flocs that are stable and have the ability to quickly reform when subjected to turbulence in pipelines, pumps and mixing in the flotation chamber. As a general rule-of-thumb for recycle units, if the carryover of redispersed flocs exceeds 50-75 mg/l in the effluent, removal efficiency is drastically lowered and excessive fouling will occur.

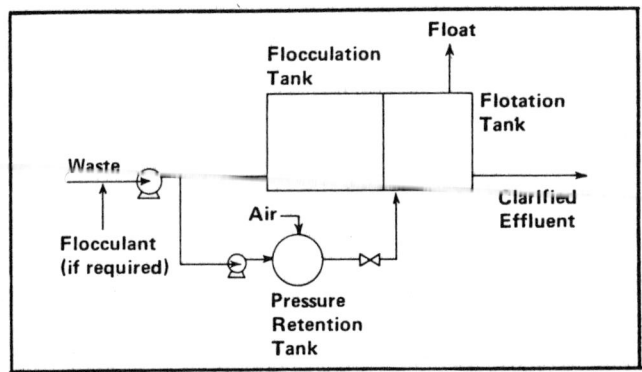

(A)

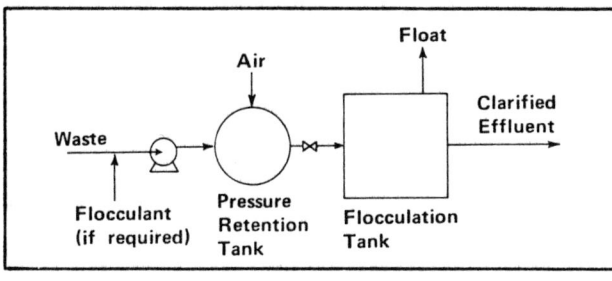

(B)

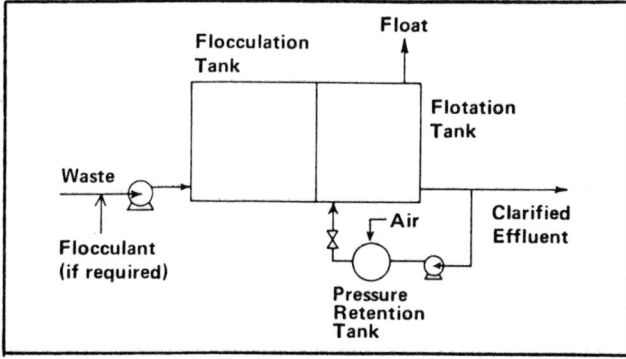

Figure 1.14 Three basic types of dissolved air flotation systems: (A) Partial Aeration; (B) Direction Aeration; (C) Effluent Recirculation.

Floated matter at the top surface of the flotation unit is removed by chain-driven flight scrapers or skimming blades. Skimmers are usually driven by variable-speed motors operated continuously, or by constant-speed motors operated intermittently. The mechanism is adjusted to remove the float at a rate established by the treatment goals.

Monitoring devices are employed to regulate air saturation in the pressurizing tank. These controls include water level sensing probes and continuous air-liquid bleed valves. The pressurized flow should be checked periodically to ensure air saturation. Often ss can interfere with the sensing devices.

The following general criteria should be incorporated in flotation system designs:

1. The flotation tank must allow aggregate rise with only minimum interference from turbulence or obstructions.
2. Effluent ports should be submerged to ensure minimal interference with the froth on the surface.
3. Obstructuve energy-dissipating devices such as baffles and walls tend to break up aggregates, resulting in reduced efficiency.
4. Turbulence or agitation in the region of the froth will result in losses of floated solids.

Detailed design information of dissolved-air flotation systems is available in the literature (9—12).

DESIGN CONSIDERATIONS FOR GRAVITY SEPARATION

Configuration

Sedimentation basins or clarifiers are rectangular, square or circular in shape and are designed for slow uniform slurry movement with a minimum of short-circuiting. They are generally classified as either horizontal flow or vertical flow systems.

The vertical flow systems are usually annular or rectangular and are wider at the top than at the base. In the annular systems the slurry is distributed at the bottom along the vessel's circumference whence it rises to peripheral or radial effluent weirs or launders. In the rectangular vessels the flow is distributed at the bottom edges of the unit whence it rises to longitudinal or transverse effluent weirs.

Horizontal units are the most widely employed designs. Rectangular designs are equipped with partitioning baffles that direct the slurry vertically to collecting basins or troughs that are positioned across and around the periphery of the tank. Sludge scrapers are pulled by chains. The scrapers transfer settled solids to a collection hopper normally near the influent side. Figure 1.15 shows a conventional design. In narrow tanks longitudinal collectors move the sludge to single or multiple hoppers at the influent end.

In the circular design the slurry enters and leaves through channels situated around the tank's periphery. The most common design has a sludge removal arm that slowly revolves about a center shaft. The removal arm or rake is normally a tapered tube with openings spaced along its length. Settled solids are collected by a slight suction action and are transported through the arm to be discharged from the bottom of the vessel near the center. Figure 1.16 illustrates a typical circular clarifier.

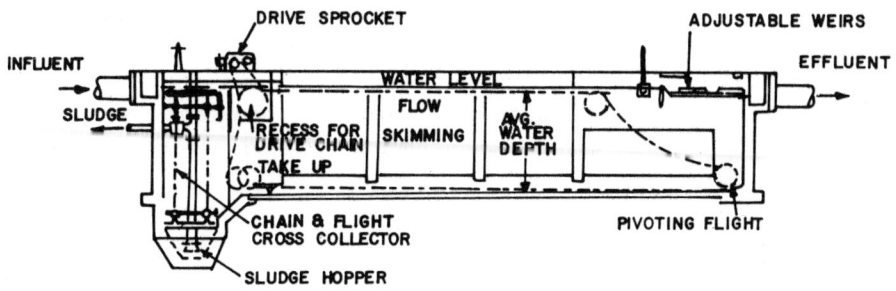

Figure 1.15 Conventional rectangular sedimentation basin.

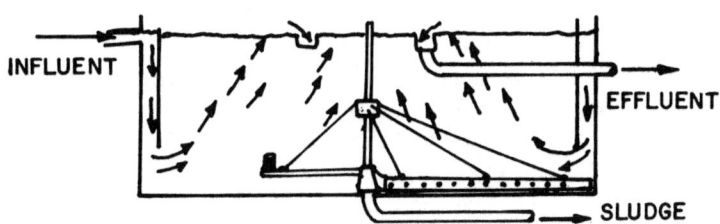

Figure 1.16 Conventional circular sedimentation basin.

Factors Effecting Efficiency

One of the major parameters governing gravity settling basin performance is the surface hydraulic loading. If the following assumptions are made, performance can be directly dependent on the surface loading (22):

1. laminar flow conditions (approaching quiescent state);
2. a uniform slurry velocity distribution;
3. discrete settling (*i.e.* no interaction between particles); and
4. no reentrainment of settled solids.

These conditions are the basis for the design criteria presented earlier; that is, all particles having settling velocities (v_y) exceeding Q/A will separate out.

In practice, there are rather large departures from the ideal conditions for gravity settling. Turbulence and "short-circuiting" can greatly alter settling performance. Short-circuiting is caused by dead spaces within the basin or by currents induced by inlet and outlet ports or by wind and density gradients. The effects are most pronounced for flocculent suspensions whose removal is governed by the detention time of the tank. High influent velocities or outlet weir rates; close placement of influent and effluent ports; or uneven heating of the slurry by sunlight all can produce short-circuiting effects. Fitch *et al.* (13) points out that short-circuiting induced by density gradients can have a very large effect on secondary settling tanks handling activated sludge mixed liquors.

Short-circuiting can be minimized in narrow, rectangular, horizontal flow designs (13). The problem is most serious in circular basins. The degree of short-circuiting in these systems varies considerably and is heavily dependent on the design of the inlet. Geinopolis *et al.* (14) have shown that inlet design considerations are more critical than effluent. The major factors to consider in minimizing short-circuiting are:

1. dissipation of influent velocity,
2. reduction of density currents associated with high suspended solids concentrations, and
3. protection of slurries from the wind or uneven solar heating.

Turbulence levels are difficult to estimate in settling basins although there are criteria for compensating for its effects (15). This is generally done by increasing the size of the tank. The amount of tank area increase will depend on the forward velocity of the liquid and with the desired removal rate.

Again, influent and effluent designs are critical. Turbulence can alter particle sizes by promoting agglomeration at the wrong time, resulting in localized settling velocities.

In rectangular designs, multiple inlet ports can distribute the flow over the width of the tank to minimize turbulence and short-circuiting. Small openings should be avoided as they can contribute to fouling. The maximum spacings of inlet ports should not exceed 10 ft, and the use of target baffles can aid in dissipating slurry velocities of the inlet jets. A manifold conduit is usually required for multiple inlets. The reader should refer to the literature (16) for information on the design of inlets and manifold conduits.

For circular designs the conventional center feed is dependent on the symmetrical baffling employed to distribute the flow uniformly in radial directions. Because of the high degree of turbulence and short-circuiting enountered with these systems, elaborate influent designs are required. For peripheral feed units inlets discharge outside a deep peripheral baffle. The flow passes underneath the baffle and enters the vessel. EPA has published a manual (27) that discusses circular designs in more detail and lists several manufacturers specializing in inlet port designs for circular clarifiers.

Other factors can greatly affect settling basin performance depending on the nature of the wastewater treated, the particular design of the system and the characteristics of the specific chemical pretreatment employed. One major problem common to both rectangular and circular types is carryover of suspended solids in the effluent. Velocity gradients within the settling basin and differences in settling rates each enhance particle contact. Both are responsible for faster moving particles overtaking slower ones. These mechanisms can greatly affect the size and settling velocity of floc and, at the same time, affect the portion of fines or unsettleable particles remaining in suspension. The attachment of smaller, unsettleable particulate onto larger particles is essential to assuring high suspended solids removal. The design must ensure that larger particles have sufficient time to agglomerate to sizes that will permit removal at the maximum surface loading conditions to the basin. This means that the detention time is a controlling parameter. The detention or residence time at a specified surface loading depends on the settling tank depth. It should be noted, however, that the rate of particle contacts due to velocity gradients increases with the forward velocity and, hence, decreases with depth.

Basin depths in this country are normally around 15 ft. The literature (13, 15) shows conflicting viewpoints as to what the recommended depth should be for designs. U.S. designs are generally deep, having low forward velocities on the order of 1 fpm at mean flow. In Europe it is common practice to design shallower systems (under 10 ft in depth) for primary settling basins with higher forward velocities (2.5—7.5 fpm) (17).

In general, an increase in the velocity gradients by agitating the inlet region of the settling basin improves performance. Common practice is to combine mechanical flocculation and settling in a single vessel. Usually compartmentation is favored to reduce short-circuiting effects. When flocculation is to be employed in upgrading existing settling basins, locating flocculation mechanisms directly in the vessel should be considered.

Table 1.2 summarizes some of the considerations the designer should take into account when sizing and selecting settling basins.

Table 1.2. General Design Considerations

1. Limit weir loadings and properly position them to minimize outlet currents.
2. Provide for uniform influent flow distribution such that inlet velocities and short-circuiting are minimized.
3. Allow sufficient sludge stor age depths to provide adequate thickening of sludge.
4. Utilize wind screens to reduce wind effects on open vessels. Wind currents altered by baffles or weirs extending above tank surfaces can cause short-circuiting.
5. Maintain equal and uniform flow in parallel systems. Equal flow distribution between basins is maintained by designing equal resistances into parallel inlet flow ports. This can be achieved by splittin the flow in symmetrical weir chambers.

Shallow Depth Sedimentation

Tube settlers have received considerable attention in the last several years, particularly with regard to municipal wastewater treatment. There are two general types of tube settling devices: horizontal and steeply inclined. These devices consist of bundles of tubes in hexagonal or square arrangements. Their hydraulic radii is around 1 in. or greater and bundle lengths are 2 ft or longer.

Inclined arrangements are usually positioned 60° or less from the ground and are fabricated in modules. These modules can be readily installed in existing settling basins. The wastewater influent enters below the tube modules and the flow travels upward through the bundles. Solid particulates move countercurrently under the influence of gravity, falling from the tube bottoms into sludge collectors. The effluent is discharged above the tube modules.

Briefly, these systems promote sedimentation in the following manner. The arrangement of multiple tubes stacked on top of each other, provides a large settling area many times that of the projection in plan of the modules. Because the hydraulic radiis of the tubes, laminar flow is readily maintained, which enhances agglomeration of particles and establishes a more uniform flow distribution. For steeply inclined designs, downward sludge movement enhances particle contact and agglomeration. A review of tube settling theory is given by Cheremisinoff (18). Viraraghavan (19) provides some background information on inclined tube settlers.

The primary advantages of these systems are as follows:

1. Since they are shallow depth-settling devices, they can improve clarification in some applications at reduced detention times over the conventional designs described.
2. Existing clarifier capacities can be augmented by as much as four times by the installation of steeply inclined modules.
3. No mechanical sludge removal provisions are required.
4. Tube bundles can be installed as an integral part of an aeration basin, thus eliminating the need for separate sludge separation provisions and recycle lines.

When installed in existing systems, an area between the inlet and tube bundles must be available to provide scum removal to prevent tube fouling. This zone should be equipped with collecting weirs. In addition, a walk area or support grid is required, and surface baffles are necessary to separate the tube settler and scum collection region (minimum basin depths are around 12 ft).

Tube fouling is a potential problem and, as such, the flow must be periodically interrupted by introducing air to remove sludge buildup. This is normally done by draining the vessel to the level of the tubes and then forcing air through them. Air is furnished either from a fixed grid or a scour system incorporated in the rake arm. Roughly 15—25 minutes is required for the effluent suspended solids concentration to return to normal after air washing. Usually, required wash cycles are done anywhere from once a week to once every 5 or 6 months, depending on the sludge carryover conditions.

REFERENCES

1. Levenspiel, O. "Chemical Reaction Engineering," 2 ed. (New York: John Wiley & Sons, Inc., (1972).
2. Cheremisinoff, P. N., and R. A. Young, Eds. "Pollution Engineering Practice Handbook" (Ann Arbor, MI: Ann Arbor Science Publishers, Inc. (1975).
3. Camp, E. "Flocculation and Flocculation Basins — theory of High Energy Mixing," Paper, No. 2722, *A.S.C.E.* (1953).
4. "Recommended Standards for Water Works," Great Lakes-Upper Mississippi River Board of State Sanitary Engineers, Health Education Service, Albany, NY (1971).
5. Cheremisinoff, P. N., and R. A. Young. "Air Pollution Control And Design Handbook — Part I" (New York: Marcel Dekker, Inc., (1977), pp. 263—280.

6. Humenick, J.J., Jr. "Water and Wastewater Treatment" (New York: Marcel Dekker, Inc., (1977).
7. "Sewage Treatment Plant Design," WPCF Manual of Practice No. 8, *A.S.C.E.* Manual No. 36 (1959).
8. Katz, W. J. "Solids Separation Using Dissolved-Air Flotation," In *Air Utilization in the Treatment of Industrial Wastes,* University of Wisconsin (1950).
9. Mulbarger, M. C., and D. D. Huffman "Mixed Liquor Solids Separation by Flotation," *J. San Eng. Div., A.S.C.E.* 96: SA4 (1970).
10. Ettelt, G. A. "Activated Sludge Thickening by Dissolved Air Flotation," *Proc. 19th Purdue Ind. Wastes Conf.* (1964).
11. Vrablik, E. R. "Fundamental Principles of Dissolved Air Flotation of Industrial Wastes," *Proc. 14th Purdue Ind. Wastes Conf.* (1959).
12. Masterson, E. M., and J. W. Pratt "Application of Pressure Flotation Principles to Process Equipment Design," in *Biological Treatment of Sewage and Industrial Wastes,* Vol. II (New York: Reinhold Pub., Co. (1958).
13. Fitch, E. B. and W. A. Lutz. "Feedwells for Density Stabilization," *J. Water Poll. Control Fed.* 32: 147 (1960).
14. Geinopolis, A., and W. J. Katz "United States Practice in Sedimentation of Sewage and Waste Solids," in *Water Quality Improvement by Physical and Chemical Processes,* (Austin, TX: University of Texas Press, (1970).
15. Camp, T. R. "Sedimentation and Design of Settling Tanks," *Trans. Am. Soc. Civil Eng.* 111: 895 (1946).
16. Naval Facilities Engineering Command, U.S. Navy "Pollution Control Systems," Civil Engineering Design Manual, DM-5, Chapter 10.
17. Kalbskopf, K. H. "European Practices in Sedimentation," in *Water Quality Improvement by Physical and Chemical Processes,* (Austin, TX: University of Texas Press, (1970).
18. Cheremisinoff, N. P. "Lamella Gravity Settler: A Compact Clarifier," *Poll. Eng.* Vol. 9: 3 (1977).
19. Viraraghaven, T. "Tube Settlers for Improved Sedimentation," *Poll. Eng.* 5 (1) (1973).
20. U.S. Environmental Protection Agency "Process Design Manual for Suspended Solids Removal," EPA625/1-75-003a (1975).

CHAPTER 2

FILTRATION AND DEWATERING

NICHOLAS P. CHEREMISINOFF
Union Camp Corp., R&D Div.
Princeton, NJ

PAUL N. CHEREMISINOFF
New Jersey Institute of Technology
Newark, NJ

PHYSICAL STRAINING

There are a variety of treatment methods available for advanced waste treatment systems. Physical straining processes is one approach in which suspended solids are removed by physical restrictions on a media having minimal thickness normal to the direction of flow.

Screening processes include rotary, static and vibrating screens, and microstrainers. Rotary screens are employed for the separation of coarse solids from a slurry stream. Figure 2.1 illustrates a rotary screen device. Basically, the waste stream enters one end of the interior of a cylindrical rotating screen and passes downward through a mesh arrangement and exits. Captured particulates exit through the opposite end of the mesh without passing through the holes in the screen. Table 2.1 lists some design data for these type systems.

Table 2.1. Design and Performance Data for Rotary Screen System[1]

	Characteristic
Percent Suspended Solids Removal	5—25
Flowrate gpm/ft^2	15—112
Wire Spacing, cm	0.025—.152
Percent Solids Treated by Weight	16—25

Static screens are units that employ a large screen surface positioned at a steep angle from the horizontal. These designs were originally devised in 1965 by the paper and pulp industry and were used to dewater and classify pulp slurries with solids contents of 6% or less. The system operates as an inclined drainage board with a screen of wedge wire construction that has opening running transverse to the flow. Basically, these systems function as follows. Raw water flows downward over

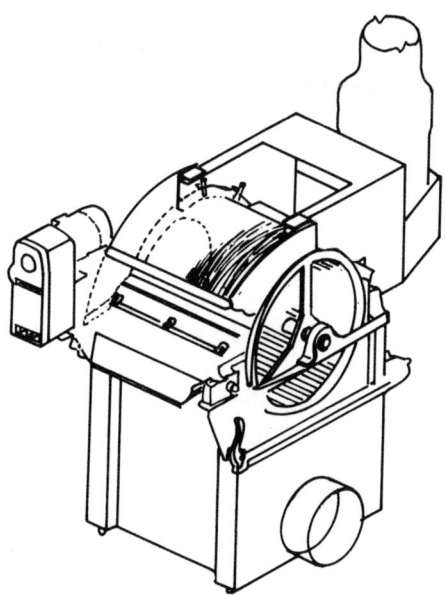

Figure 2.1. Rotary screen design (1).

the face of the screen and tends to flow through the upper third of the screen. Dewatered solids simultaneously travel down the remaining portion of the screen into a sump. Static screens require little maintenance provided there are not large amounts of fats or oily solids present in the slurry.

Usually the system has a headbox that runs across the width of the unit. On the Hydrasieve by C-E Bauer (Division of Combustion Engineering Co.), the headbox consists of a lightweight hinged baffle at the upper portion of the screen. This design tends to reduce flow turbulence in the unit (1).

Inclined static screens are usually constructed of stainless steel. Lightweight units are constructed of fiberglass housings and froms which are roughly 25% less expensive than the all steel designs (2).

Inclined units have been satisfactorily operated at loading capacities ranging from 4—16 gpm in. of screen width. Hydraulic capacity is a function of the fluid viscosity, temperature, the solids loading and screen spacing (*i.e.,* spacings of individual slots). The optimum slot width is determined from the screen's capacity per foot of width, which can be computed from empirical correlations derived from bench scale studies.

It should be noted that inclined screens have considerably lower suspended solids removal efficiencies than sedimentation basins; however, they have received favorable acceptance primarily because they are capable of removing bulky materials that may potentially foul subsequent sludge handling units. Size openings limit their ability to remove fine grit; as such, separate grit removal units are usually installed after inclined screen equipment.

Vibrating screens are usually designed as rectangular or circular structures that support wire mesh screens. The raw water is directly introduced onto the screen surface. Suspended solids collect on the screen surface, while the water passes downward through the screen. The trapped solids are removed by a vibrating mechanism, which causes them to move to the outer periphery of the screen element where they are removed for disposal or further treatment. Grease, fats and stringy solids present problems in that they tend to hang on the wire mesh and cause binding.

Microstrainers (also called microscreeners) consist of a drum which is closed at one end and revolves on a horizontal axis at peripheral speeds up to 100 fpm. The drum is sheathed with either stainless steel or a plastic fabric of special weave. The fabric may have perforations as small as 23 μ (i.e., approximately 144,000 openings/ft^2 fabric). Usually a tank surrounds the drum. The tank is sectioned into two compartments. One compartment is designated for the influent slurry and the other for the strained effluent. Raw water to be treated enters the open end of the drum and moves radially through the fabric face of the drum. Suspended solids are trapped on to the inside surface of the drum fabric. The effluent stream containing strained solids normally returns to the primary settling basin for recycling. The amount of effluent used for recycling is approximately 5% of the total microstrained influent. Slime and algae buildup are major problems with this design; however, this can usually be controlled by shining a high-intensity ultraviolet light on the fabric surface.

Wastewater loading rates for these systems range from 2.5–10 gpm/ft^2; however, this can vary with the type of fabric used. Treatment facilities in Lebanan, OH and Chicago, IL, report using a 23-μ filter fabric that showed an average suspended solids removal of 80% in activated sludge effluent. Final effluent suspended solids were roughly 6–8 mg/l and BOD removal around 40–50%, with final effluent BOD in the range of 9–12 mg/l. The backwash rate was 3–5% of the total influent, which contained roughly 700–1,000 mg/l suspended solids (3).

Careful analysis and characterization of suspended solids in slurries to be treated must be made. Of particular interest are the concentration and the degree of flocculation, as these parameters can greatly affect microstraining efficiency as well as capacity and backwashing requirements (4). Note that provisions for backwashing and supplemental cleaning facilities are an essential part of the design for maintaining good design capacity. Table 2.2 lists typical operating conditions for microscreening units.

Boucher (5) correlated the effects of influent solids properties with flow capacities of different fabrics. Boucher assumed that for steady-state laminar flow, the head loss through any specified strainer fabric increases exponentially with the

Table 2.2. Operating Conditions and Parameters for Microscreens[1]

Backwash Flow and Pressure	— 2% of throughput at 50 psi
	5% of throughput at 15 psi
Peripheral Drum Speed	— 15 fpm at 3 in. head-loss
	125–150 fpm at 6 in. head loss
Head Loss Through Screen	— 3–6 in.
Hydraulic Loading	— 5–10 gpm/ft^2 of submerged surface area
Submergence	— 66% of drum surface area

volume passing through per unit area. This can be expressed as follows:

$$\frac{\Delta P}{A P_o} = e^{IV} \tag{1}$$

where ΔP = pressure drop across the screen, ft)
       A = submerged screen area ft^2
       P_o = pressure on the influent side of the screen ft
       V = slurry volume flowing per unit area ft^3/ft^2
       I = filterability index, ft^{-1}

A hydraulic capacity expression for the continuous operation of a rotating drum microscreen is given by Equation 2 (6):

$$v = \ln\left[\left(\frac{\Delta P}{F}\right)\left(\frac{I\epsilon}{R}\right) + 1\right]\left(\frac{R}{I\epsilon}\right) \tag{2}$$

where v = mean flow velocity through the submerged screen area, fps, and is defined as Q/A
       Q = total flow through the microscreen, ft^3/s
       F = fabric resistance coefficient, ft/fps
       ϵ = decimal fraction of screen area submerged
       R = drum rotational speed, rpm

Note that $\Delta P/F$ is the initial flow velocity through the clean screen as it enters submergence. The fabric resistance coefficient, F, depends on the specific screen fabric and varies inversely with mesh size.

Equation 2 is shown graphically in Figure 2.2. The graph is a plot of v versus $\Delta P/F$ with $I\epsilon/R$ as a parameter. Lines of constant flow ratio, E, are shown where the constant flow ratio is defined as the ratio of the mean velocity through the screen to the initial velocity when the screen enters submergence (E = v/($\Delta P/F$)). Mixon (6) recommends choosing an $I\epsilon/R$ such that E ≤ 0.5; above this operating

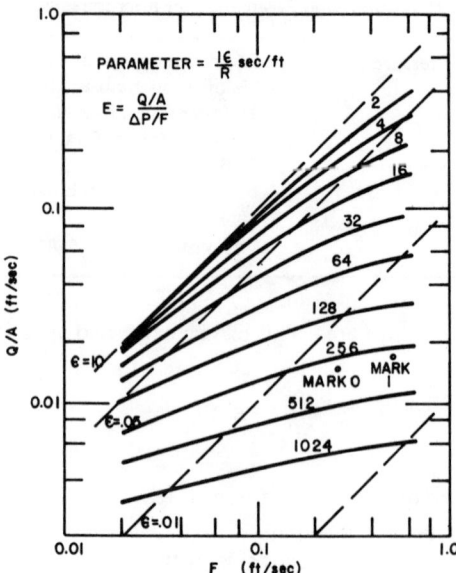

Figure 2.2. Microscreen capacity chart (3).

limit it is difficult to form a mat of trapped solids on the drum. The filterability index must be determined from test data obtained from pilot microscreen prototypes. Similarly performance characteristics must be evaluated on the basis of pilot studies.

Some general recommendations for the design and operation of microscreens are as follows:

1. Microscreens can be used in place of clarifiers as a polishing unit for low flowrate effluent from trickling filters. This is generally the case for low solids concentrations in a well-flocculated stream.
2. Suspended solids removal tends to increase slightly at lower hydraulic loadings.
3. Shearing action on influent should be kept to a minimum. High shearing can cause flocculated particles to break up (7).
4. Lower drum speeds tend to produce better quality effluent (8) because better straining action develops because of a thicker mat of solids, which tends to build up at lower speeds.
5. Select an operating submergence level in the range of 70–75% of the drum diameter. The operating drum submergence level is a function of the effluent slurry level and head loss through the mat. The minimum submergence is the level of the slurry inside the drum when there is no flow over the effluent weir; the maximum level is established by a bypass weir, which allows flows in excess of unit capacity to be bypassed.
6. At maximum submergency, the maximum drum differential should not exceed 15 in.
7. The top effluent weir should be positioned at the specified submergency level.

FILTRATION

The most common piece of equipment used in water treatment is the rapid sand or gravity filter. Figure 2.3 shows a cutaway view of a rapid sand filter bed. A bed of sand approximately 2 ft deep lies over a graded gravel layer. A network of underdrains runs under the gravel bed. Usually the support structure is concrete and a typical depth is about 9–10 ft. Raw water is distributed over the sand layer and is drawn through the filter bed by suction from the floor of the basin and water pressure from above.

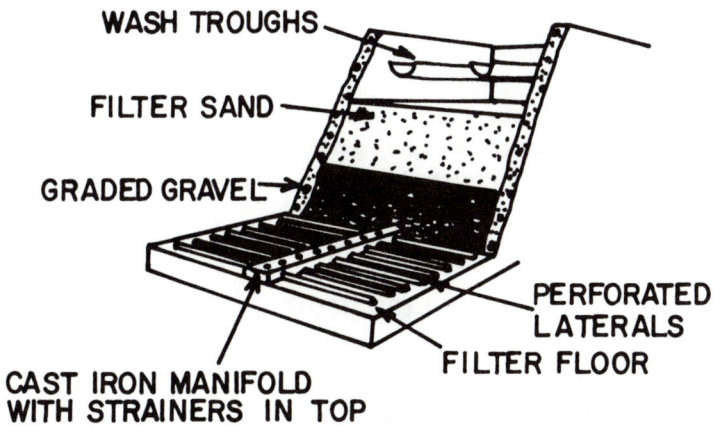

Figure 2.3. Cross-sectional view of a sand filter.

Mechanisms controlling filtration, even in the simple sand filter, are complex. Straining, flocculation and sedimentation each play a role in removing suspended solids by this method. Slurries to be treated by gravity filtration must first be chemically treated and large flocs removed. As such, most filtration units are positioned after sedimentation units. Carryover of coagulants from sedimentation treatment is necessary for gravity filters to function properly and remove microscopic particulates that would otherwise pass through the bed. Excess carryover can cause heavy mats to form on the filter surface and thus clog the bed. Improperly coagulated streams may also channel through the bed and be flushed out before being trapped. This generally produces a highly turbulent effluent. Ideal filtration conditions occur when nonsettleable coagulated flocs become entrapped in the pores of the bed. This results in a condition known as "in-depth" filtration.

Ideal filter media should have the following characteristics:

1. The media should be sufficiently coarse to provide sizeable pore openings for retaining large amounts of floc
2. It should also be sufficiently fine to entrap suspended solids
3. It should have adequate depth to provide long filter runs.

When the media becomes exhausted, it is cleaned by backwashing. Wash troughs are usually positioned above the media's top surface to collect backwash water and carry it away from the filter basin.

The first filter media used was sand. Beds contained sand particles of uniform size in the range of 0.4–0.5 mm. After the backwashing operation, particles become redistributed in such a manner that the finest are on the top and the largest grains on the bottom. When this occurs, the greatest removal occurs near the upper portion of the bed. As such, many of the early designs had low removal efficiencies. Early sand systems were also difficult to clean. Various heavy impurities (such as mud balls or flocs) tended to redeposit in the gravel underlayer when the bed expanded during backwashing, thus causing further clogging problems. If higher backwash rates were attempted, sand carryover iin the washwater resulted.

Modern units employ surface wash agitators, which consist of a rotating arm positioned above the sand bed and beneath the wash trough. These washing aids are equipped with spray nozzles that direct a mist at right angles to the bed surface. During backwashing the bed expands, and water is introduced to the rotating agitator. The result in increased turbulence providing much improved cleaning action.

FILTER MEDIA

Mixed media beds are often preferred to sand beds. One common mixture is coal, sand and garnet, with garnet being the finest medium. Normally, the three materials are sized so that after backwashing a nearly homogeneous media develops. Without stratification of filter media the bed more closely approximates the concept of a uniform decrease in pore size with increasing bed depth, although in some designs. stratification is preferred.

The use of diatomaceous earth as a filter has been limited to swimming pools and portable field units. It has been applied to the clarification of secondary effluents on a pilot scale; however, no fullscale operations have been reported in the literature. Diatomaceous earth is capable of producing a high-quality effluent but not in a range of solids loadings encountered in municipal or industrial wastewater treatment.

Diatomaceous earth filler induces filtration by making use of a thin layer of precoat that develops around a proous septum and strains out suspended solids in the slurry as it passes through the filter cake and septum. Vacuum from the effluent side or head pressure from the influent provides the driving force. Head loss through the medium increases to a maximum due to the deposited suspended solids. Once this maximum head loss is reached, the bed is cleaned by backwashing and the filtration process initiated once again. In pilot studies for secondary effluent treatment relatively large quantities of diatomaceous earth were necessary to maintain a continuous feed operation and to prevent rapid solids buildup resulting in large head losses.

There are various grades of diatomaceous earth commercially available. In general, the coarser the grade, the greater its permeability and solids-handling capacity. On the other hand, the finer grades result in a higher-quality effluent.

The most common arrangement with this type of filter consists of a support structure that is a fine metal screen, a porous ceramic or a synthetic fabric. During filtration, the slurry to be filtered is precoated to a predetermined head loss. During the precoat step water containing diatomite runs through the filter and deposits a mat on the screen. The body feed consists of the slurry stream with a small amount of diatomaceous earth mixed in. The body feed is drawn through the filter produced by the initial water throughput. Common loading rates for these units are 1–5 gpm/ft^2.

The controlling factor in secondary effluent treatment with diatomaceous earth filters is the amount of body feed required during the run time (run time begins when feed is introduced to the filter and terminates when a limiting head loss is achieved). Body feed rate is the largest contributor to operating costs and affects the cycle time between backwashings; this, in turn, has an effect on the installed filtering area. A detailed discussion of design criteria for diatomaceous earth filters is given by Bell (8).

FILTRATION UNITS

A general description of rapid sand filters has already been given. The influent that flows down through the media has a surface loading in the range of 2–6 gpm/ft^2. When the pressure drop reaches 8–10 ft, the filter is taken out of service and backwashed (downtime). The backwashing step can either be a reversal of flow using clean effluent, or by the use of cleaning aids as previously described. Another common technique is to first scour the media in an upflow direction with air (this dislodges particulates from the media) and next flush the media with effluent water. The amount of filtered effluent used for backwashing is roughly 2.5% of the total flow volume. This effluent, with a high solids loading, can be returned to the primary settling tank for treatment (3).

The removal of suspended solids by rapid sand filtration depends on several factors: (1) particle size and density, (2) filter media surface area, (3) bed depth and porosity, (4) filtration rate, and (5) fluid properties such as viscosity and surface tension.

Small particles, up to 3 μ in size, deposit through the filter surface by diffusion. Larger particles are removed by sedimentation.

Solids removal efficiency in these units can be enhanced (particularly small particulates) by the use of coagulants in the rapid mix basin or by the use of a polymer coagulant that can be directly deposited on the filter surface prior to passing the effluent onto the bed. This approach has been successful in removing color, odor and some viruses from wastewater streams.

Rapid sand filters are most efficient after suspended solids have been flocculated and large particles removed. Under these conditions only colloidal particles reach the filter bed, resulting in a smaller volume of particles being deposited in the bed. The end result is longer run time. BOD removals have been reported as high as 99% with 12—18-in. depth media with loading rates of 4—5 gpm/ft^2.

Another variation of rapid sand filters is an arrangement utilizing upflow filtration, which employs coarse-to-fine gradation of media particle size. Retention bars at the top of the bed prevent carryover and bed fluidization. These systems are operated at loading rates of 3 gpm/ft^2 with 6—20 ft of head loss. Removal efficiencies of 85% suspended solids have been achieved. Figure 2.4 shows still another rapid sand filter design that employs only a single-graded media. This type of system is employed in tertiary wastewater treatment.

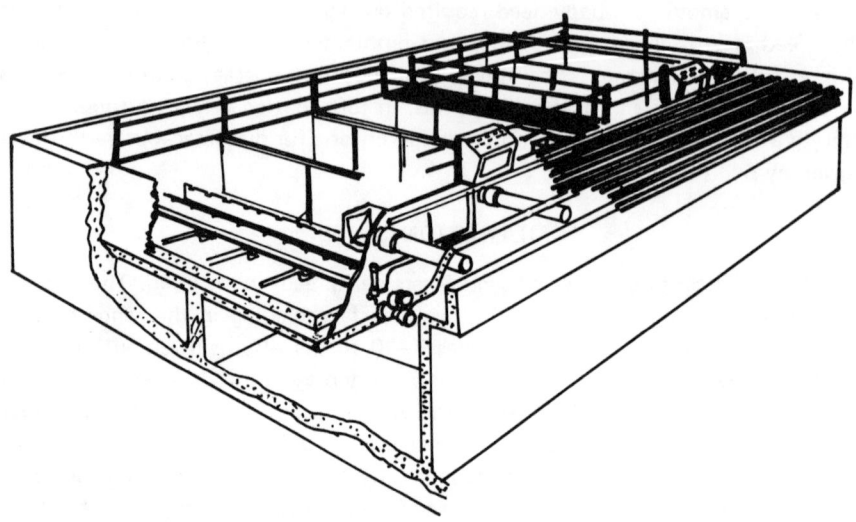

Figure 2.4. Regenerative surface single-graded rapid sand filter designed for tertiary wastewater treatment of high solids-bearing slurries (Hydro-Clear Corp.)

Slow Sand Filters — As the name implies, these units handle slow influent flows, roughly 3 gal/hr/ft^2. Wastewater introduced to the bed percolates down through the filter media until the head loss becomes excessive. When this occurs the influent starts to accumulate above the top surface of the bed. When the water level reaches the top wall of the filter assembly, the unit is drained and dried out. The surface layer of sludge formed is removed manually as these units are designed without backwash capabilities. Usually two or more units are maintained to allow continuous operation.

Pilot studies in England show that slow sand filtration of suspended solids can provide removal efficiencies of 61—63% and BOD removals in the range of 37—43%

with flowrates of 1.5–2.5 gal/hr/ft^2 and influent solids of 10–16 mg/l. These systems are subject to rapid clogging and often require shutdown during cold weather conditions.

Moving Bed Filters — These are continuous, rapid, sand-type filters that are designed for countercurrent sand and influent flow direction. Sand is forced through a cone by a diaphragm. During the diaphragm's pushing stroke sand is forced from the surface of the cone down into a hopper where it undergoes cleaning. The influent flows through the sand bed in a countercurrent direction to the sand. This produces a continuously moving surface layer by exposing clean filter media. When the diaphragm relaxes regenerated sand moves into the void remaining in front of the diaphragm. The main advantage of this design is that it eliminates the need for downtime backwashing.

Removal rates are in the range of 80–90% suspended solids and 60–75% BOD removal. Flowrates and solids loadings are generally significantly higher than the conventional fixed bed rapid sand filters.

Coagulation of the influent stream with alum and/or polymer coagulants is required to prevent extensive shearing of the secondary treatment plant biological floc.

In-Depth Filters — These systems employ multimedia with a gradation of grain size from coarse to fine in the direction of flow. The most widely used multimedia filter is the trimedia, usually composed of a 7-in. layer of coal (graded coarse to fine), which covers a layer of coarse to fine sand, which in turn covers a bed of garnet. This three-layer bed lies over a bed of gravel covering the underdrain system.

The coal bed depth may vary from 8–22 in. with grain sizes of 1–2 mm; silica sand depth varies from 9–12 in. with grain sizes of 0.6–0.8 mm; and the garnet bed depth varies from 3–8 in. with grain sizes of 0.4–0.8 mm. Both grain sizes and bed depths depend on the nature of the effluent and its flowrate, as well as the suspended solids loading and characteristics.

With this type of design, it is desirable to have well-defined or stratified layers. Some intermixing occurs during the initial startup; however, after sufficient run time this will cease, resulting in an almost continuous gradation of filter media from coarse to fine. These designs can handle an upper loading limit of 120 mg/l at 5 gpm/ft^2 for 15–24 hours (final head loss on the order of 15 ft). An intermediate settling tank is usually installed to absorb system shocks, which will ensure stable loading. Removal efficiency can be enhanced by the use of chemical coagulation prior to filtration and/or by the addition of polymers directly to the filter bed.

Greater influent loading and longer filter runs have been obtained by housing the mixed media in a pressure vessel. This produces higher available heads (up to 20 ft). Pressure vessel systems have two distinct advantages: (1) the filter effluent can flow downstream without further pumping to the next stage in the treatment operation; and (2) since longer filter runs can be achieved, less washwater is required. Backwashing of gravity and pressure mixed media filters is done in the same manner as for rapid sand filters.

Grass Plots — Grasslands have been employed as filtration systems for secondary or packaged treatment plant effluents. The effluent is distributed over the land through an arrangement of channels. The effluent flows from these channels over land-planted grass that are of the deep-rooted variety, and down into a series of secondary channels.

A number of these installations in England have been proven to work reasonably well. Suspended solids removal is in the range of 61–75%, with final effluent values of 12–16 mg/l. BOD removal is in the range of 55–57%, with final effluent concentrations of 54–58 mg/l. Average flowrates reported are between 0.2 and 0.7 gal/hr/ft^2. In winter operation, ice forms at the surface and flow is continued below.

Plate and Frame Filters — Plate and frame filtration is primarily a batch process. It is used mainly for filtration of small quantities of high solid slurries, particularly in the chemical, dyestuff and sludge handling industries. Driving pressures in these units typically range from 50 to as high as 1,000 psi. Plate and frame filtration is a labor-intensive operation. These units operate by depositing the cake in a separate frame while the filter media is supported over the plate. The more common design used today is a recessed plate, which eliminates the need for the frame. Large recessed plate units utilize synthetic medias such as polypropylene, polyester, nylon and saran as support media for establishing the cake. These units are capable of containing up to 80 ft^2 of filtration area per chamber. Depending on the filtration characteristics of the sludge being processed, some units can handle up to 8 ft^3 of cake per chamber. Modern units may have as many as 120 chambers of large recessed plates in a plate stack resulting in packaged filter areas of 9,600 ft^2 (and 960 ft^3 of cake per discharge).

With large plates there is no need to employ manpower to move plates as was practiced with the older plate and frame filter designs. Mechanized filter skeletons are employed, which allow discharge of these units in less than 15 minutes.

Plate-type pressure filtration units are designed to handle between 1.5 and 8% dry weight solids in the influent. Feed conditioning by chemical flocculation, polyelectrolyte or admixing with filtration improving aids is required. Positive displacement pumps, either of the diaphragm type, progressing cavity or ram type, are employed in feeding pressure filtration units.

Vacuum Filtration — There are four basic types of vacuum filters: drum, belt, coil and continuous disc filters. Rotary drum vacuum filters are the most commonly employed of the three for dewatering sludge. Basically, a cylindrical drum covered with synthetic cloth media rotates partially submerged in a vat containing the chemically treated waste. A scraper or "doctor blade" discharges the cake from the drum. This is usually supplemented by a pressure blowback in the discharge zone. Air pressure is supplied from inside the drum just ahead of the doctor blade to aid in sludge removal from the filter medium surface. The filter medium is never removed from the drum during operation.

The filter medium on the belt vacuum filter is a continuous belt that leaves the drum surface and moves over a short-radius discharge roll.

This action causes a sharp change in direction and causes the sludge cake to discharge from the drum surface. Some designs also supplement this by using a doctor blade. The belt is washed on both sides prior to returning to the drum.

The third type uses two layers of stainless steel coil springs arranged in a corduroy fashion about the drum. Figure 2.5 shows a typical coil filter design. The layered springs serve as the filter medium. At the end of the dewatering step, the double layer of springs leave the drum and are separated from each other. This separation causes the filtered sludge cake to be lifted off the lower coil springs and is discharged from the upper layer by a positioned tine bar. Both springs are washed separately by an arrangement of sprayers and reapplied to the drum by means of grooves aligning rolls.

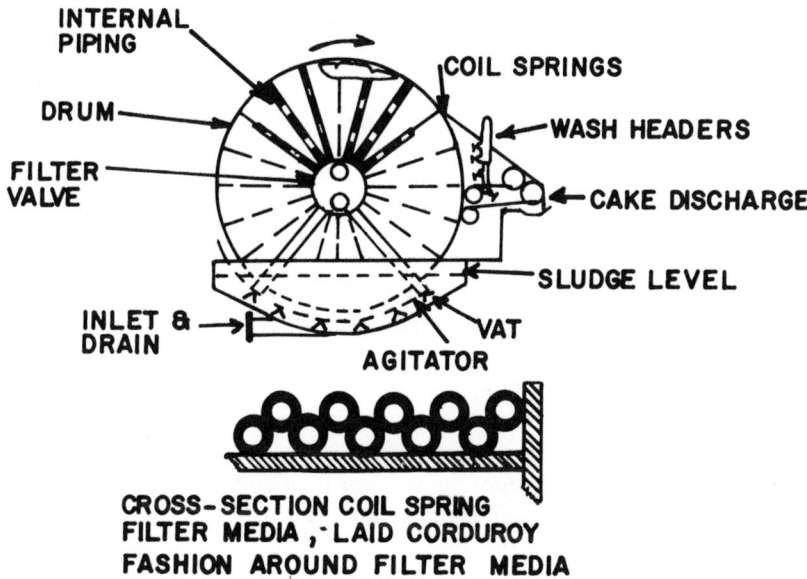

Figure 2.5 Rotary vacuum coil filter.

Continuous disc filters utilize filtering surfaces on both sides of several disc mounted perpendicular to a rotating horizontal center shaft. Each disc collects filtrate and scrapers are used to discharge the cake. Figure 2.6 illustrates its operation.

For all the designs between 10 and 40% of the drum surface is submerged in a vat containing sludge. The submerged surface area is referred to as the cake-forming zone. Vacuum applied to the submerged section causes filtrate to pass through the media and the cake begins to build up. In the rotating drum design, as it rotates, each section passes through the cake-forming zone and then to the cake drying or

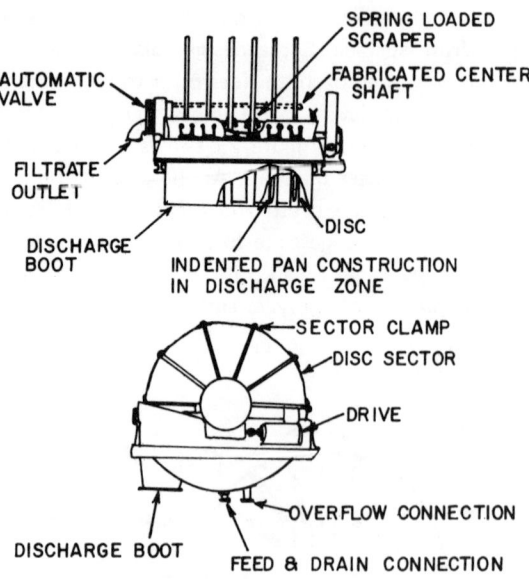

Figure 2.6. Continuous-disc vacuum filter

dewatering zone. The dewatering zone is also under vacuum and begins when the drum carries formed cake out of the sludge vat. This zone represents 40–60% of the drum surface and ends when the vacuum is shut off to each section as it enters the cake-removal zone.

Auxiliary equipment employed with all four designs consists of a filtrate receiver, a vacuum pump and a filtrate pump. Lines leading from the filter rotary valve transfer filtrate and air from the vacuum filter to the filtrate receiver where they are separated. The filtrate pump draws the filtrate off the bottom of the receiver while air is evacuated from the top of the receiver via the vacuum pump.

For biological sludges, additional pretreatment or sludge conditioning equipment is required. This may include high temperature and pressure wet oxidation or chemical conditioning (such as ferric chloride, polyeletrolytes or a combination).

Chemical metering pumps, chemical solution or day tanks may also be required. Depending on the quantities and nature of chemical treatment required, lime slaking equipment, polyelectrolyte wetting, aging equipment and storage facilities may also be necessary.

The rate of vacuum filtration is expressed in units of pounds of dry solids removed per hour per square foot of filter area (known as the yield). As a rule of thumb, expected filter yield of chemically conditioned sludge is around 1 lb/ft^2/hr for each percentage of solids. In general, yield increases at a continuous rate with increasing chemical dosage pretreatment. When conditioning is not sufficient enough to flocculate the finer suspended solids, filtrate solids become high. It is essential to operate vacuum filtration units in regions of low filtrate solids to prevent returning excessive quantities of solids back to the various treatment stages. Final cake moisture tends to decrease with increasing chemical pretreatment as well as increasing solids concentration in the influent.

The ultimate size of a vacuum filter unit will depend on the desired dry solids production, specified yield, which depends on the thickeners and nature of the sludge, and on the operating time.

Sludge samples should be acquired prior to the final design and tested for filterability in the laboratory. The most common testing method is the Buchner funnel test. Basically, this involves pouring a sample of chemically conditioned sludge into a filter-paper-lined funnel and dewatering under vacuum. This allows prediction of the suitability of sludges for filtration as well as a test of pretreatment chemicals for the particular sludges considered. It is primarily a screening test in that little scaleup information can be obtained.

A more extensive test involves the use of a filter leaf with roughly 0.1 ft^2 of effective area attached to the vacuum apparatus. The filter medium should have characteristics similar to the full-scale unit under consideration. The unit is inserted upside down into the slurry sample and suction is applied, thus simulating the cake-formation zone. The unit is then withdrawn and dewatered for a time period that is proportional to the actual drum rotational speed. The cake formed is removed from the leaf face and filtrate is drawn into a vacuum flask where it can be tested for solids content. Information obtained from this type of study can be applied to actual scaleup design.

Ultrafiltration — This is a form of membrane separation employing relatively coarse membrane separation at low pressures. The method uses a thin semipermeable polymeric membrane. Smith *et al.* (9) report that ultrafiltration is successful in removing suspended solids and large-molecule colloidal particulates in the range of 0.002–10.0 μ. By regulating pore sizes, fluid transport and suspended solids retention are achieved. Smith reports that in applications with activated sludge-ultrafiltration suspended solids, BOD and bacteria are removed to nearly 100%.

The membrane area of these systems depends on the flux. The flux is determined by the membrane construction and fouling characteristics of the slurry. Due to surface fouling, membrane flux decreases with time. Flux-reducing effects can be reduced by removing some of the physical foulants such as any organic acids that might be in the waste stream.

It is desirable to design these sytems for maximum membrane area, to increase fluxes. The membrane is usually constructed from two layers. The surface layer is an extremely thin homogeneous polymer (roughly 5 μ in thickness). The second layer is for surface support and is an open cell of 5–10 mil thickness. The membrane is supported on a porous sheet, usually paper, for added mechanical strength. The thin surface controls transport and rejection properties of the membrane.

Membrane porosity is characterized by water permeability; however, this does not resemble the stabilized long-term flux on a process slurry. Bemberis et al. (10) note that fluxes are typically 7—10 gal/day/ft² of membrane surface area.

Pressure is the driving force for transporting water through the membrane. Pressure gradients of roughly 25 psi is required for normal operation. In general, total system pressure does not exceed 50 psi.

CENTRIFUGATION

Centrifuges separate suspended solids out of solution by applying centrifugal force. There are three basic designs: (1) the imperforated basket type, (2) the solid bowl type, and (3) the disc nozzle type.

The first centrifuges developed were of the basket type shown in Figure 2.7. Basically, these systems consist of a spinning cylinder that generates high centrifugal forces that force suspended solids against the drum wall. The influent enters from the bottom center and clarified effluent is discharged over a lip ring at the top. When the cake in the bowl reaches a specified height, operation is halted and the collected sediment is removed with a skimmer or knife plow. Feed is again resumed after sludge removal.

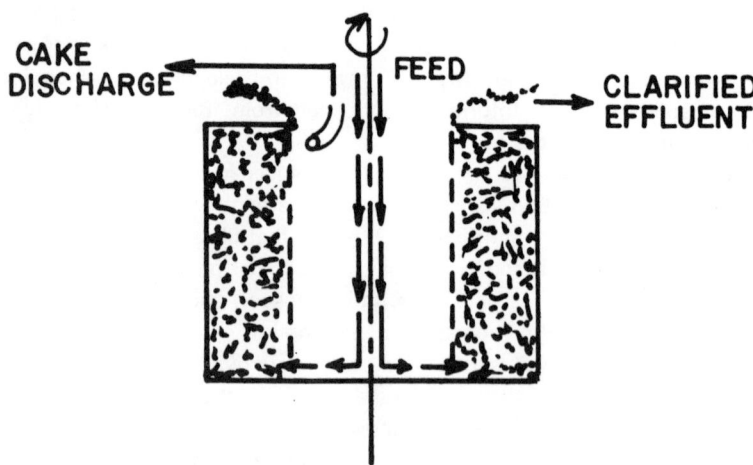

Figure 2.7 Solid-bowl basket centrifuge.

The solid bowl type is shown in Figure 2.8. In this unit solids build up as a cake against rapidly rotating sides while liquid is drained off by flowing over the top of the basket. Both systems described can operate in either batch or continuous mode.

Centrifuges are employed as either dewatering or classifying systems in tertiary treatment plants. They are also utilized in recausticizing systems in paper mills where sludge is washed and thickened in a mud washer and then fed to a centrifuge

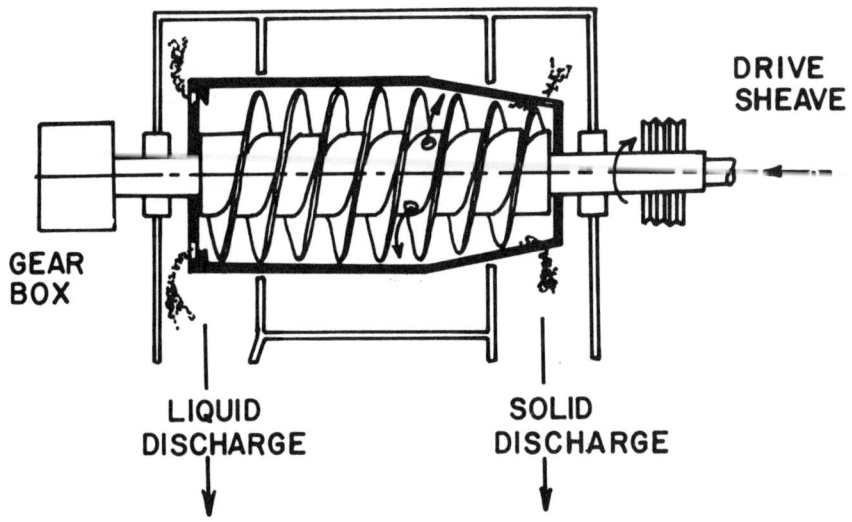

Figure 2.8. Solid-bowl continuous-discharge centrifuge.

at 25–35% solids. The cake product produced is usually in the range of 65–70% solids.

The rotating assembly of the disc-nozzle type centrifuge consists of a stack of conical discs. Incoming feed to the unit is distributed between a multitude of narrow channels arranged as spaces between the stacked conical discs. Settleable solids fall down through the liquid layer flowing in the channels to the underside of the disc forming the top of the channel. Solids slide down the underside of the disc to a sludge compaction zone. Thickened sludge discharges through nozzles into a sludge collection section in the casing. A portion of the thickened sludge is usually recycled through the sludge compaction zone to achieve higher solids concentrations. The discharge of sludge through the nozzle orifices sets an upper limit on the size of particles that can be handled efficiently by the system. Sludge pretreatment is required before centrifugation. The disc nozzle centrifuge is capacity limited to thickening of waste-activated sludge; however, it has high solids removal efficiencies (11).

All three types of centrifuges operate by the same principle — that is, suspended solids are removed from suspension by imposing a centrifugal field that is roughly 1,000–6,000 times the force of gravity. The fundamental differences are the methods described above by which solids are collected and discharged from each unit. The method of discharge establishes the size and nature of particulates that can be handled suitably in each design. Table 2.3 lists the type of sludges that can be suitably handled in the three types discussed.

In general, centrifuges have received wide acceptance in municipal and industrial

Table 2.3. Sludges Handled by Different Centrifuge Units

Solid-Bowl Type	Imperforated Basket	Disc-Nozzle
Ground Screenings	WAS dewatering	Thickening waste
Raw Primary	Aerobic digested	activated
Primary Digested	Alum treated	
Combined Raw Primary/WAS	Industrial wastes	
Heat Treated	Thickening WAS	
Lime Treated		
Alum Treated		
Pure Oxygen		
Thickening WAS		
Industrial Wastes		
Classification		
Mixed Digested		

waste treatment applications. Centrifuges have the following advantages:

1. They are relatively easy to operate and maintain.
2. They have low to moderate space requirements.
3. Odor control can be built into an enclosed design.
4. They are relatively easy to start up and shut down.
5. Units display good flexibility in handling daily variations in sludge characteristics.

The actual selection of a specific centrifuge design depends on the characteristics of the sludge to be dewatered, hydraulic loading rates, influent solids concentration and the ultimate solids removal desired by the overall plant design.

Design and Performance

As shown in Table 2.3, the solid bowl (conical design) is the most versatile of the three and is most often preferred. This design can be directly compared with a horizontal clarifier because both processes are applications of Stoke's law:

$$v = \frac{2R^2 (s-s^1)_g}{9\mu} \tag{3}$$

where v = settling or separation rate
R = mean or hydraulic radius of the particle
s and
s^1 = specific gravities of the settleable particle and the fluid, respectively
μ = viscosity of the fluid
g = gravitational constant

The major difference between the two processes is that in centrifugation, the g factor may be 1,000–6,000 g. This, of course, means greater settling rates and higher solids removal. Centrifugation is more efficient for solids removal than sedimentation for one other reason — solids only have to settle over a very short distance, a matter of inches, whereas in a sedimentation basin, particles have to settle through several feet.

The major design variables that must be evaluated before selecting a specific centrifuge design for a particular dewatering application are: (1) bowl length, (2) bowl diameter, (3) speed, and (4) scroll speed differential.

Retention time increases with bowl length. This results in drier cakes and clearer centrates. It should be noted, however, that the baseline economics must be carefully evaluated since power requirements will also increase. Electrical load requirements are proportional to the volume contained in the centrifuge.

Increasing the centrifuge diameter has the same effect as increasing the bowl length — that is, they both lead to drier cakes and clearer centrates at the same feed rate. In addition to higher power demands that will be imposed, more critical balancing will be necessary since the weight mass moves farther from the center in large-diameter centrifuges.

Higher speeds also produce drier cakes and clearer centrates. However, this also leads to greater wear on system components, which affects maintenance schedules and cost. Low-speed centrifuges must be preferred for handling abrasive materials.

Systems are normally equipped with a conveyor arrangement. As conveyor speeds are increased, the centrifuge's throughput will increase but at the expense of a wetter cake and dirtier centrate. By altering pulleys and gear boxes on existing systems, bowl and scroll speed differential can be adjusted for improved performance (12). During operation, sludge flow, the degree of chemical conditioning and pool depth sshould be monitored and balanced to maintain optimal operation.

Centrifuge throughput will increase as sludge flow is increased; however, as noted, a penalty must be paid in concentrate clarity. High flows can cause flooding. If flowrates are too low, extremely dry cakes result, which results in excessive wear on the conveyor scroll. An optimum flow must be determined that will allow sludge to be pumped at the highest rate without producing an excessive accumulation of fines in the concentrate.

Deeper pools can produce higher quality centrates but weaker cakes. Shallow pool depths lead to drier cakes but more turbid centrates. In many designs the pool depth can be altered externally while in operation. It is worthwhile to examine a range of pool depths during the initial startup period of a new installation.

Chemical conditionining parameters may be more difficult and time-consuming to control. Those parameters of particular interest are dosage, dilution and injection station. Cylinder tests are recommended for basing proper polymer selection and optimum dosage. Exact dosages should be evaluated on the basis of centrifuge tests. If a unit has been designed with more than one polymer injection point, all stations should be evaluated to determine the optimum effect. Internal polymer feed points are designed to introduce the polymer after the heavy, readily settleable solids have been separated out of solution. It should be noted, however, that it is often desirable to add polymers to the total sludge stream. In this case, the heavier solids aid in bridging and flocculating the finer particulates. In general, the conditioning time is a relatively short period; therefore, it is essential to dilute polymers to a level that will ensure good distribution throughout the slurry without overloading the centrifuge with dilution water.

Performance criteria should be well outlined to effectively evaluate a program. Dewatering performance is usually defined in terms of solids recovery, *i.e.*, the fraction or percentage of suspended solids in the sludge influent that forms the discharge cake. Recovery can be calculated from the following expression:

$$\eta = \frac{(t-d)(f-c)}{(f-d)(t-c)} \qquad (4)$$

where η = percent solids recovered
  f = total solids in the feed stream, expressed as weight percent on total slurry weight
  c = weight percent of total solids in the concentrate
  d = weight percent of total dissolved solids
  t = weight percent of total solids in the sludge cake

DEWATERING WITH POLYMERS

There are a variety of methods available for concentrating or dewatering sludges; however, many of them are complemented by the use of chemicals to improve their speed and effectiveness. The traditional chemicals employed were metal salts such as alum or iron, usually in conjunction with lime. Organic polymers have recently replaced the use of metal salts in many applications.

In general, colloidal particles can be removed from suspension by any one or a combination of the following mechanisms: (1) destabilization, (2) microfloc formation, (3) agglomeration, or (4) physical entrapment.

For colloidal particles to coalesce, interparticle repulsion forces must be neutralized (see Section VII). This is known as destabilization. Certain cationic chemicals can be employed to decrease the net negative charge of suspended particles.

Once charges have been stabilized, particle collision occurs and particles can form small microflocs through chemical bridging, hydrogen bonding and Van der Waal's forces. If mild agitation is applied, microflocs will combine to form macroflocs.

Not all particles will undergo destabilization, microfloc formation and agglomeration. A fraction of the suspended solids will be removed by physical entrapment in macroflocs already formed.

As already mentioned, inorganic coagulants are the traditional chemicals used to achieve coagulation and agglomeration. These inorganics are the di- and trivalent metal salts (aluminum, calcium and iron). When calcium salts enter water they form calcium ions. When the aluminum or iron salts are added, they form trivalent metal complexes with water. These complexes are referred to as polyaluminum hydrates or polyferric hydrates. These complexes are cationic in nature and are capable of stabilizing colloidal suspensions. In addition, they have sufficient chain length to bridge between particles.

Betonite clays and activated silica are other examples of inorganic coagulants. Betonite clay behaves like a nucleus for floc formation and also functions as a weighting agent. Activated silica is primarily a flocculant that aids in macrofloc formation, which has been initiated by other inorganic coagulants (13).

The more widely accepted chemical coagulants today, the organic polyelectrolytes, are either natural or synthetic. Natural starches and gums have been employed for a number of years as flocculants. Their primary function is to serve as a bridging agent in much the same manner as activated silica.

Synthetic organic polyelectrolytes have received wide acceptance because they can be used at significantly lower dosages than conventional inorganics and/or natural coagulants for many applications. More often they are used as aids with inorganic coagulants.

The organic polyelectrolytes have high molecular weights. They are essentially long-chain molecules composed of a number of repeating units that are held together by covalent chemical bonds. Usually, some or all of these units can be dissolved or ionized in water solution.

The synthetic organic polyelectrolytes fall into three categories: nonionic, anionic and cationic. An example of a nonionic polymer is polyacrylamide. Examples of anionic polymers are the hydrolyzed polyacrylamides and copolymers of acrylamide and acrylic acid. Anionic polymers exist as long-chain molecules in solution. Nonionic polyacrylamide is in the form of random coils in solution.

For relatively stable suspensions, strong cationic polyelectrolytes are required to destabilize the suspension. Homopolymers of dimethyl diallyl ammonium chloride and ethylenimine are most effective in this case.

Batch laboratory tests should be performed to evaluate the most effective and economical chemicals that should be employed in treating. The standard jar testing system contains four to six mixing paddles synchronized by a common drive. In this manner four to six different treatment schemes can be tested and compared to each other under the same conditions. Experiments should be designed to vary the degree of mixing, mixing time and settling times. Graduated cylinder tests can also be used for comparing various chemical treatments.

Streaming current detectors and zeta meters are more elaborate testing methods. These methods measure surface charge and the products ability to alter charge; however, theey do not provide a measure of the degree of floc formation, floc settling ability or the supernatent turbidity as the conventional jet tests.

Kemmer et al. (14) provides a detailed account of the chemistry of complex polyelectrolytes in wastewater treatment. Also, more detailed information on various polymers that can be used as coagulants is available in the literature (15—17).

REFERENCES

1. U.S. Environmental Protection Agency, "Process Design Manual for Suspended Solids Removal," EPA 625/1-75-003a (January 1975).
2. Ginaven, M. E. "The Hydrasieve — A New Simplified Solids-Liquid Separator," *Paper Trade J.* (January 1970).
3. Cheremisinoff, P. N., and S. M. Feller, "Wastewater Solids Separation," *Poll. Eng.* 7 (8) (1975).
4. Diaper, E. W. J. "Tertiary Treatment by Microstraining," *Water Sew. Works* 116 (June 1969).
5. Boucher, P. L. "A New Measure of the Filterability of Fluids with Applications to Water Engineering," *I.C.E. J.* (British) 24:4 (1947).
6. Mixon, F. O. "Filteraabilitty Index and Microscreener Design," *J. Water Poll. Control Fed.* 42 (November 1970).
7. Hydrocyclonics Corp. *Rotostrainer Bulletin.*
8. Bell, G. R. "Design Criteria for Diatonite Filters," *J. Am. Water Works Assoc.* 54 (October 1962).
9. Smith, C. V., Jr., D. Di Gregario and R. M. Talcott "The Use of Ultrafiltration Membranes for Activated Sludge Separation," Purdue Indiana Waste Conference May 7, 1969, Water Renovation of Municipal Effluents by Reverse Osmosis, Water Quality Office, USEPA Project EPA 17040 EOR (February 1972).
10. Bemberis, I., Hubbards and Leonard. "Membrane Sewage Treatment Systems," American Society of Agriculture Engineers, Winter Meeting 1971.
11. Cheremisinoff, P. N., J. Beardsley, F. Biermann, A. Blok, W. Hallen, J. Innes, R. Kormanik, J. Lodholz, L. Minnick, R. Moll, W. Schwoyer and C. Wall. "Sludge Handling and Disposal: A Special Report," *Poll. Eng.* 8 (1) (1976).
12. "An Introduction to Centrifuges," Nalco Chemical Co. Bulletin TF92.
13. Reilly, P. B. "Water and Wastewater Treatment for Removal of Suspended Solids," Bulletin 12-65, Polymer Department, Calgon Corp.
14. Kemmer, F. N., and K. Odland. "Chemical Treatment," Nalco Chemical Co., *Chem. Eng.* Reprint No. 175 (October 1968).
15. Hart, J. A. "On Improving Wastewater Quality," Nalco Chemical Co., *Water Sew. Works.* Reprint No. 192 (September 1970).
16. Herner, B. P. "Polymer Applications Increasing in Water and Waste Treatment," *Water Poll. Control* (July 1975).
17. Lo Sasso, R. A. "Polymers Can Help," *Water Wastes Eng.* (June 1973).

CHAPTER 3

VACUUM FILTRATION

LEONARD P. EGEE
Komline-Sanderson Engineering Corporation
Peapack, NJ

INTRODUCTION

Most modern industry is dependent upon some type of liquid-solid separation process, and many separations are accomplished by vacuum filtration. This form of filtration is defined as the separation of a solid from a liquid by causing the liquid to pass through a permeable medium which retains the solids. The driving force for separation is created by the vacuum applied to the filter, resulting in a pressure across the filter medium exerting up to 14 lb/in.^{2} to produce the filtration. In most industrial applications the driving force is limited, for practical reasons, to 10–12 lb/in^2.

This chapter reviews equipment associated with vacuum filtration. The discussion will not include a detailed theoretical analysis of filtration nor will it cover filter media or filter aids, since these topics are discussed elsewhere.

Vacuum filters are available in a variety of types and can be either continuous or batch operated. The type of filter employed for a specific application depends to a large extent on the physical characteristics of the suspension being separated. Such factors as suspended solids content and particle size are obviously the controlling factors and influence not only the filtration rate but the type of filter that can be employed.

The primary factor affecting the actual rate of cake formation for a given filtration area is the specific cake resistance, R, which is a physical constant for the material being filtered. Specific resistance may be defined as a factor numerically equal to the pressure differential (driving force) required to produce a unit rate of filtrate flow through a unit weight of cake. A general filtration equation may be written as follows:

$$Y = \frac{\sqrt{2 \Delta P}}{\mu R T C}$$

where Y = yield as expressed in time per unit volume per given filtration area lb/ft^2/hr
  ΔP = driving force psi
  μ = filtrate viscosity centepoises
  R − specific resistance sec^2/g

T = filtration time sec
C = slurry concentration, %

When driving force, viscosity, specific resistance and slurry concentrations are held constant, the filtration rate varies with time. When filtration rate is plotted against time on log-log paper, a straight line having a negative slope of 0.5 usually results.

For continuous filtration applications, the rate at which a cake is formed normally determines the type of vacuum filtration equipment that should be used for an application. Figure 3.1 shows this relationship. The chart is divided into five areas, which show areas of applicability for the various types of equipment to be discussed. If the basic cake formation properties under vacuum are known, Figure 3.1 will aid in filter selection and will allow the reader an understanding of the cake-forming time and thickness requirements for the various types of commercially available filters.

Before discussing the various vacuum filter types, certain basic common denominators may be described:

Filter Media: Most modern-day filters employ synthetic filter cloth, precoat material, woven wire cloth or stainless steel springs as filter medium.

Drainage Member: The filter medium is normally supported over a relatively open area so that vacuum can be evenly applied and filtrate removed without causing a significant pressure drop. The supporting structure is generally called a "drainage member."

Filtrate Outlet: The pipes or channels that remove filtrate and air have their inlet under the drainage member and terminate at the point of discharge from the filter.

Vacuum Receiver: When filtrate and air are transported from the filter in combined form, the mixed flow of liquid and gas must be separated to allow efficient removal of each phase. A vacuum receiver is used for this and normally consists of a tank large enough to reduce the flow velocity and also acts as a surge chamber for the accumulation of liquid and air volumes, which can then be removed by appropriate means.

Vacuum Pump: A mechanical gas pump or eductor, which produces a negative pressure on its inlet side.

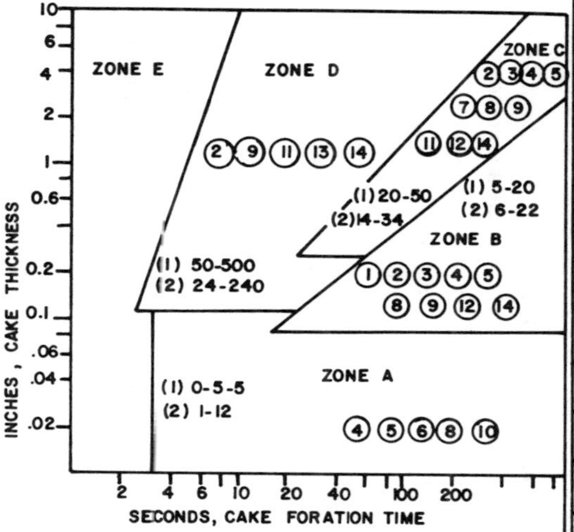

Type of Filter	Cont. Opera-tion	Cont. Media Wash	Thin Cake Disch.	Cake Wash Capab.	Wash Separ-ation	Hydr. Capac.	Filtrate Clarity	Relative Horse Power	Relative Cost	Vapor Retain. Design
1 LEAF	NO	NO	NO	YES	YES	MED.	GOOD	MED.	LOW	NO
2 NEUTSCHE	NO	NO	NO	YES	YES	LOW	EXC.	LOW	LOW	YES
3 • SCRAPER DISCHARGE	YES	NO	NO	YES	YES	MED.	GOOD	MED.	MED.-LOW	YES
4 • STRING DISCHARGE	YES	NO	YES	YES	YES	MED.	GOOD	MED.	MED.-LOW	YES
5 • ROLL DISCHARGE	YES	NO	YES	YES	YES	MED.	GOOD	LOW	MED.-LOW	YES
6 • PRECOAT DISCHARGE	YES	NO	YES	YES	NO	MED.	EXC.	MED.	MED.	YES
7 • COILSPRING FILTER	YES	YES	NO	NO	NO	MED.	FAIR	MED.	MED.-HIGH	NO
8 • ENDLESS CLOTH	YES	YES	YES	YES	YES	MED.	GOOD	MED.	MED.	YES
9 SINGLE CELL ROTARY DRUM	YES	NO	YES	YES	YES	HIGH	GOOD	HIGH	HIGH	YES
10 TROMMEL	YES	NO	YES	NO	NO	LOW	EXC.	MED.	MED.-HIGH	NO
11 • TOP FEED	YES	NO	NO	NO	NO	HIGH	FAIR	HIGH	MED.-HIGH	YES
12 ROTARY DISC	YES	NO	NO	NO	NO	HIGH	FAIR	MED.	MED.-LOW	NO
13 HORIZONTAL ROTATING PAN	YES	YES	NO	YES	YES	HIGH	GOOD	HIGH	HIGH	YES
14 HORIZONTAL ENDLESS CLOTH	YES	YES	YES	YES	YES	HIGH	GOOD	HIGH	MED.-HIGH	YES

• MULTI-COMPARTMENT ROTARY DRUM

—THE CHART MAY BE USED TO PREDICT THE TYPES OF FILTERS WHICH MAY BE APPLICABLE FOR A SPECIFIC FILTRATION APPLICATION IF BASIC CAKE FORMATION CHARACTERISTICS ARE KNOWN.

—THE NUMBERS SHOWN IN VARIOUS ZONES CORRESPOND TO FILTER TYPES SPECIFIED IN FIRST VERTICAL COLUMN. SOME CHARACTERISTICS OF EACH FILTER TYPE ARE LISTED IN THE VERTICAL COLUMNS TO THE RIGHT OF THE FILTER TYPE.

(1) POUNDS DRY SOLIDS/HR/FT²
(2) GALLONS FILTRATE/HR/FT²

Zone E: (1) 50-500, (2) 24-240 — ② ⑨ ⑪ ⑬ ⑭
Zone D: (1) 20-50, (2) 14-34 — ⑪ ⑫ ⑭
Zone C: (1) 5-20, (2) 6-22 — ② ③ ④ ⑤ ⑦ ⑧ ⑨
Zone B: ① ② ③ ④ ⑤ ⑧ ⑨ ⑫ ⑭
Zone A: (1) 0-5-5, (2) 1-12 — ④ ⑤ ⑥ ⑧ ⑩

X-axis: SECONDS, CAKE FORMATION TIME (2, 4, 6, 10, 20, 40, 100, 200)
Y-axis: INCHES, CAKE THICKNESS (.02, .04, .06, 0.1, 0.2, 0.6, 1, 2, 4, 6, 10)

Figure 3.1

Continuous vacuum filters generally have the ability to produce greater yield rates than other types of filters, despite the fact that the driving force for filtration is less than that available in pressure-type filters and centrifuges. The reason for increased yield is the short residence time and relatively thin cake formation available on the equipment. While cake dryness is not always equal to that which can be achieved with alternate processes, more uniform cake drying and cake washing (where needed) are obtain obtained.

Because of the lower driving force used on vacuum filters, the associated design problems are much less severe than those encountered when dealing with high pressures or centrifugal force. This simplicity carries over to the operation of the equipment and manifests itself in the ability of vacuum filtration equipment to operate without difficulty on a continuous basis. By far, the greatest number of applications for vacuum filters is on continuous processes; however, batch filters have found use in some areas.

BATCH FILTRATION

The two vacuum filter types described below represent commercially available units which are relatively inexpensive to install. They offer an advantage when process conditions change frequently, causing the need for variable filtration or cake-washing times. Since they operate on a batch basis, adequate time can be provided for washing of the filter medium so as to diminish the possibility of blinding the medium.

Vacuum Leaf Filter

The best known leaf type of vacuum filter is the Moore filter, which consists of a group of leaves manifolded together and connected to a vacuum system. The leaves are made up of rectangular-shaped drainage members, which are enclosed by the filter medium. The groups of leaves are transported, generally by an overhead crane, along a path that immerses them successively in a feed-slurry tank, a wash-liquor tank and a cake-receiving container. In many cases, air blowback is used to aid cake discharge. The Moore filter is very simple in design and allows flexibility in the cake-forming and washing cycles. It has the disadvantages of being labor intensive, requiring large floor space and being subject to the danger of cake falling off the leaves during drying or transport.

Vacuum Nutsche

The vacuum Nutsche is generally a cylindrical vessel divided into compartments by a horizontal filter plate, which supports the filter medium. Vacuum is applied to the lower compartment; filtrate is either accumulated in the lower chamber or may be discharged continuously. The cake is removed manually. The vacuum Nutsche has the advantage of being simple to build and can be constructed of any material;

it has the disadvantages of laborious cake removal and limited throughputs.

CONTINUOUS VACUUM FILTERS

Continuous filters are those that operate essentially without interruption, in that the material to be filtered can be fed continuously and its solid and liquid components discharged continuously. There are four types of continuous vacuum filters: rotary drum, rotary disc, rotary horizontal and horizontal belt or "endless cloth" types.

The most commonly used continuous vacuum filter is the multicompartment rotary drum, which will be discussed first, along with its various modifications and adaptations. Patented in England in 1872 by William and James Hart, the principles of this filter have changed little, if at all, from those described by this extraordinary patent. This type of filter was introduced to the United States in 1908 by E. L. Oliver.

Multicompartment drum filters range in size from 10 ft^2 (0.9 m^2) to over 1,000 ft^2 (93 m^2). The most common sizes are shown in Figure 3.2.

										FACE										
		1'	2'	3'	4'	5'	6'	7'	8'	9'	10'	11'	12'	13'	14'	15'	16'	17'	18'	19'	20'
DIAMETER	3'	9.4	18.8	28.2	37.7	47.1	56.5														
	4½'		—	42.4	56.5	70.6	84.8	98.9	113	127											
	6'				75.5	94.2	113	132	151	170	188	207	226								
	8'						150	176	201	226	251	277	302	327	352	377	402				
	10'								—	283	314	345	377	409	440	471	502	534	565	596	628
	12'										377	415	452	490	528	565	603	641	679	716	754

Figure 3.2

The rotary drum vacuum filter consists of a horizontally positioned cylindrical drum supported on trunnions and partially submerged in a slurry reservoir containing the material to be filtered. The outer circumference of the drum is divided into a number of shallow longitudinal sections by means of "division strips;" each section can be considered an individual vacuum chamber, isolated from the adjoining sections and communicating through piping to a central outlet located at one of the end trunnions. Figure 2.3 shows the interior of a large drum with section pipes leading to a manifold and terminating at the filtrate trunnion. As a general rule, the filter designer attempts to limit the pressure drop (from the drum deck to the filter valve) to not more than 1 in. Hg at full flow conditions. Variables include the depth of the drainage member, the number and size of drainage pipes, and the size and shape of the filter valve. Small-diameter pipes restrict flow and limit hydraulic capacity, while large-diameter filtrate pipes require very large trunnions. Piping to each section is generally a combination of "lead" and "lag" pipes. Lag pipes drain a

Figure 3.3 Filter drum interior shows internal construction and piping. (Komline-Sanderson).

section emerging from the slurry, whereas lead pipes are requied to drain a section that has passed the 12 o'clock position and is rotating downward. Lead pipes are particularly useful in applications involving large volumes of cake wash. The basic principles of operation of the rotary drum vacuum filter are shown in Figure 3.4.

The filtration cycle on a rotary drum vacuum filter starts when a section of the rotating drum becomes submerged in the slurry to be filtered. Vacuum is applied to the drum section through the section-and-header pipes by means of a bridging arrangement in the stationary valve on the end of the filtrate trunnion. Vacuum is maintained on that section, and on each section following, as it is submerged, creating a continous filtration zone. Filtrate is drawn through the medium and solids are deposited on the surface as the drum rotates. The cake formation soon becomes its own filter medium, with the installed filter medium acting as a cake support. Cake continues to form as a function of time until the section leaves the slurry and is exposed to the atmosphere. Air is drawn through the cake, forcing out the mother liquor. This part of the cycle is known as the "drying cycle." Cake formation normally uses 25—35% of the drum circumference, with corresponding drying zones of approximately 60—55%.

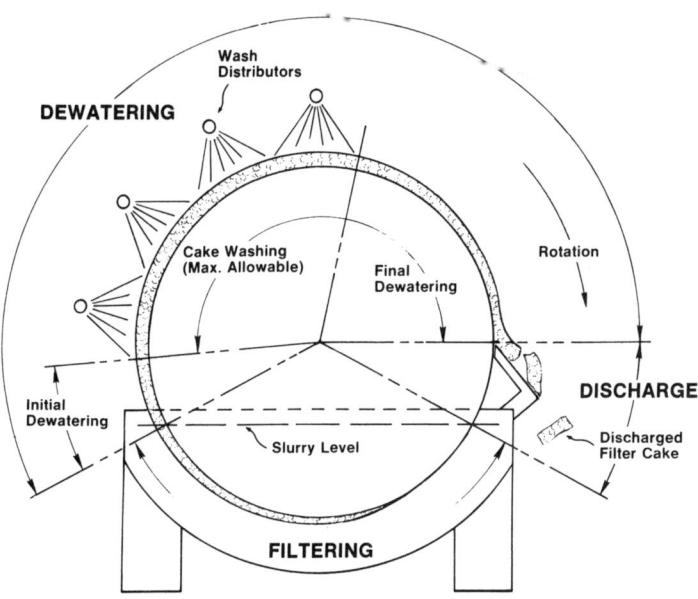

Figure 3.4 Basic principle of continuous rotary vacuum filtration.

In many applications it may be desirable to remove impurities from the filter cake by washing it with another liquid. Washing may be performed on the drum immediately following its emergence from the slurry. The wash liquid, generally water, may be introduced to the surface of the cake by wash header pipes or drip pans. When full coverage wash zones are required, a "traveling wash blanket" may be employed to prevent erosion of the cake by the wash liquor. The blanket rides on the cake and prevents direct impingement of the wash liquor while providing a constant and uniform flow of wash liquor over the entire cake surface.

In the final zone, the vacuum is released as the drum section enters the discharge zone, allowing the cake to be removed from the surface of the filter medium by one of several methods which will be discussed under types of discharges.

In some applications compression rolls may be used to mechanically compact the cake and lower the cake moisture beyond that achieved by vacuum alone. Multiple numbers of rolls are often used to allow gradual compression. Rolls may consist of either weighted discs or a single solid roll. Pneumatic or hydraulic loading may be used to supplement the roll weight. Figure 3.5 shows a compression and wash assembly that includes two compression rolls and a traveling wash blanket.

The heart of the filter is the filter valve, which provides the means for applying vacuum during the cake-forming, washing and dewatering portions of the cycles. Figure 3.6 shows a typical valve. Bridge blocks within the annular ring of the valve

Chapter 3. Leonard P. Egee

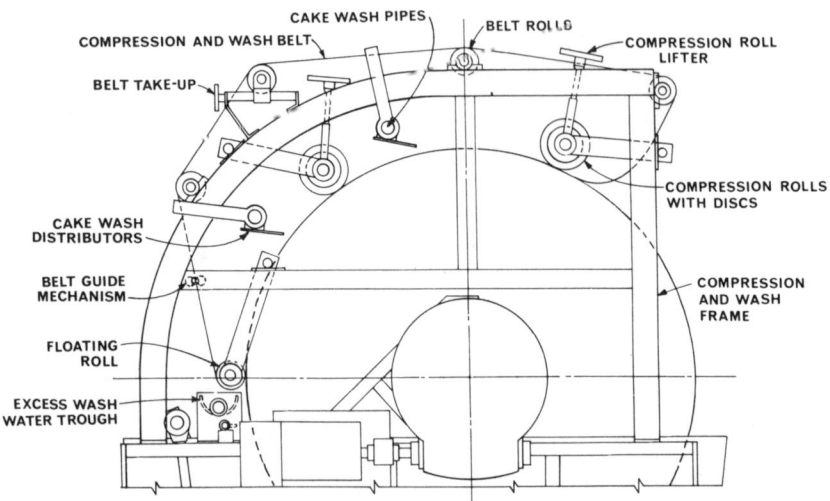

Figure 3.5 Compression and wash assembly, showing two compression rolls and traveling wash blanket. (Komline-Sanderson).

Figure 3.6 Automatic filter valve showing adjustable bridge blocks and polypropylene wear plante. (Komline-Sanderson).

are used to sectionalize the valve to control the vacuum level in those sections that require it, while releasing the vacuum or applying blowback to the sections in the discharge area. These blocks may also be used to regulate the vacuum between sections or to facilitate the separation of mother liquor from wash liquor when required.

The filter medium, normally a synthetic cloth, but in some instances a wire mesh or precoat, must be supported to allow drainage of liquor and passage of vapor. Drainage support grids may be made of wood, metal or plastic, and are designed to allow adequate room for free drainage between the filter medium and the drum deck. Figure 3.7 shows a typical snap-in plastic grid. For most applications, the grid is between ¾ in. (19 mm) and 1 in. (25 mm) deep, and can handle moderate amounts of filtrate. When operating on free-filtering materials, the sections are made deeper to allow free drainage and, in some instances, may be as deep as 2 in. (50 mm).

Figure 3.7. Plastic drainage grid. (Komline-Sanderson)

In most operations the drum revolves continuously at a speed of 2—10 minutes per revolution; on some very free-draining materials, drum speed may be increased to 0.3—0.5 minutes per revolution.

Drums are driven from one end through a gear or chain drive and, in most instances, are supplied with an adjustable-speed drive that allows a speed range of approximately 10:1. The drive is customarily mounted on the end opposite the filter valve.

Chapter 3. Leonard P Egee

The most common filter vat is designed for 37½% submergence of the drum, as this is the highest submergence obtainable without resorting to trunnion stuffing boxes. If the filter capacity is limited by cake-forming ability, it is possible to operate at a submergence of 55% or more; however, the design (as mentioned above) is complicated by the requirement for stuffing boxes. Consideration must also be given to the position of the discharge, as high submergence may lead to a discharge height that will not allow satisfactory discharge of the filter cake.

Drum submergence is considered to be one of the three basic operating variables; the other two being drum speed and vacuum level. The maximum capacity of a drum filter is reached when it is operating at the fastest drum speed and highest vacuum and submergence that will allow a suitably dry and dischargable cake. In general, drum speed is the principal rate control. The vat level, which controls submergence, may be fixed by overflowing slurry back to the feed tank or by a level sensing device that controls the slurry feed line. If the vat slurry level is to be changed, the valve bridge blocks must be changed to satisfy the new conditions.

The outlets of the filtrate valve are connected to a filtrate receiver, which is a cylindrical tank that functions as a separting device. All continuous facuum filters, with the possible exception of the single-compartment filter, require separators. With fast-filtering slurries, even the single-compartment-type vacuum filter requires receivers because of the large volume of air that must be handled.

For normal operations the receiver is of simple design: a single tank with a side inlet through which the liquid and vapor mixture enter, an outlet in the upper head through which the vapor is removed and a bottom outlet for the liquid filtrate. The liquid can be removed either by a centrifugal pump, or, if height allows, by a barometric leg. To obtain good separation between the vapor and liquid, the receiver is designed to limit the cross-sectional velocity to less than 3 ft/sec.

One receiver is usually adequate; however, in some instances, it may be required to separate a wash liquor from the mother liquor. This is done with two receivers. In some cases, filtration results may be improved by use of different form-and-wash and drying vacuums. If so, a two-receiver system with vacuum regulation is used. Figures 3.8–3.11 show some of the more common receiver arrangements.

Many applications require some form of a hood to prevent the escape of objectionable vapors, to prevent contamination of the material being filtered or to confine spray from wash headers. A typical vapor-retaining hood is shown in Figure 3.12. This type of hood is relatively simple and may be constructed of light metal or fiberglass. It is open to the atmosphere and has a top outlet that is connected to a suction fan for removal of unwanted vapors.

Some processes require a hood that will prevent oxidation of the product, loss of solvent vapors or the escape of flammable or toxic gases. These applications require a filter with a vapor-tight hood. Such a filter is very expensive to build, as it must be completely sealed and gasketed. The discharge area is especially complicated as it is generally required to be vapor tight. Figure 3.13 shows a vapor-tight filter with a filter cake discharge screw incorporating an air lock.

The principal advantages and disadvantages of multicompartment rotary drum vacuum filters are:

Advantages:

1. continuous in nature; very low labor requirements;
2. flexible; can handle a wide range of materials;
3. maintenance costs very low; and
4. cake washing can be effectively performed;

Disadvantages:

1. requirement for installation and maintenance of a vacuum system;
2. high initial cost; and
3. limitation on vapor pressure with hot or volatile liquids.

Cake Discharge

A number of different discharge arrangements may be used to remove the cake from a rotary drum vacuum filter. The type selected depends on the nature of the material being filtered.

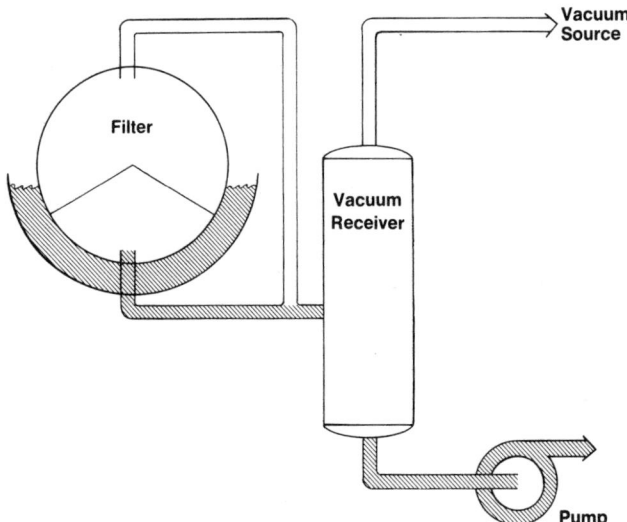

Figure 3.8 The most universal design is that of the single receiver where the same vacuum is employed for cake formation and drying.

Scraper Discharge

The first method, and still the most widely used, is the knife or scraper discharge. This design incorporates a blade that removes the cake from the drum by direct contact with the filter cake. A low pressure air blowback (Figure 3.14) is

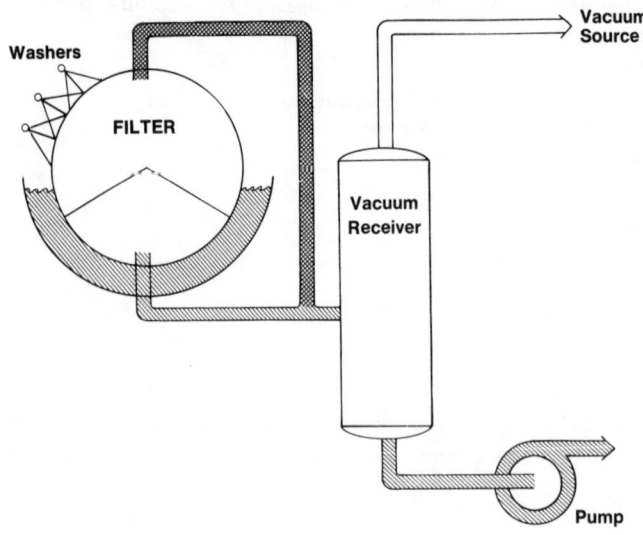

Figure 3.9 The single-receiver system may be employed when other functions are included in the filtration cycle, such as the addition of a cake wash system.

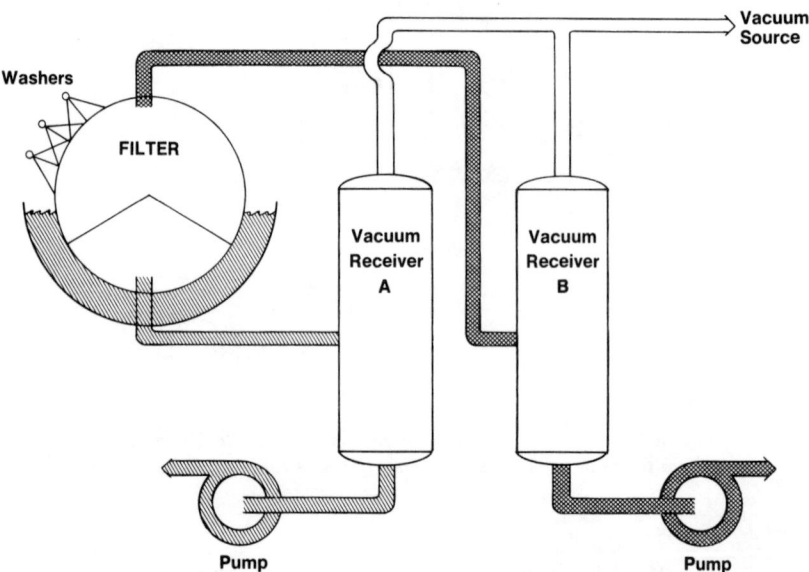

Figure 3.10 This design includes a two-receiver system with the same vacuum for cake formation, washing, and drying phases of the cycle. The wash liquor and mother liquor are separated as shown.

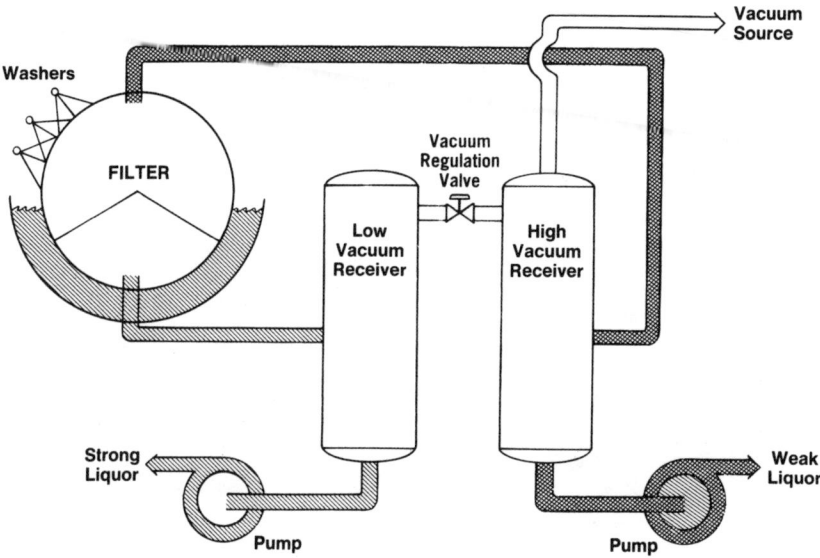

Figure 3.11 The two-receiver system may utilize differential vacuums for cake formation, washing, and drying, as well as separation of the wash liquor and mother liquor. This system is shown, however, modifications including countercurrent washing systems may be incorporated into the design if required.

Figure 3.12 Rotary drum vacuum filter with vapor-retaining hood. (Komline-Sanderson).

Chapter 3. Leonard P. Egee

Figure 3.13 Rotary drum vacuum filter with vapor-tight hood. (Komline-Sanderson).

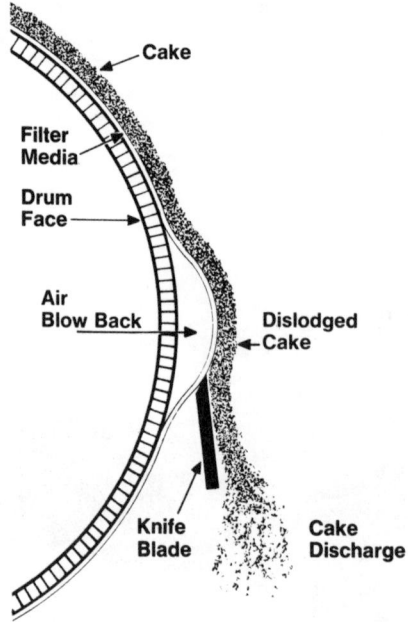

Figure 3.14 Scraper discharge.

sometimes used to assist cake separation from the medium. The filter medium (synthetic or wire cloth) is secured to the drum by caulking it into division strips by banding or by wire winding. The scraper discharge is normally used on granular or crystalline materials where filter medium blinding is not a problem and cake thickness is in excess of ¼ in. (6 mm).

String Discharge

After the development and introduction of the scraper discharge unit, it was found that with the application of the blowback air applied at discharge to remove the cake, moisture was blown back into the cake, thus rewetting the cake. This moisture is from the small droplets clinging to the internal piping that cannot be completely drained during the drying cycle. The string discharge was developed to solve this problem. String discharge filter design involves the use of parallel strings tied completely around the filter, passing over the discharge and return rolls. The string, generally spaced out about ¼ in. (6 mm) to 3/8 in. (9 mm) apart, leave the drum on a tangential plane and lift the filter cake from the filter medium. As they pass around the discharge roll, the cake is separated from the strings and discharged. The string discharge design is sometimes recommended for effective handling of gelatinous or other types of cohesive cakes.

Roll Discharge

After the development and introduction of the string discharge type of filter, it was found that there was still a need for a unit to discharge tacky types of cakes, which neither the scraper nor the string discharge could handle effectively. The roll discharge (Figure 3.15) satisfies this need.

This design employs a roll, often rubber covered, at the discharge point of the filter. The roll rotates at a slightly faster peripheral speed than the filter drum and is in contact with the cake that has been formed on the filter medium. The cake is transferred from the filter drum to the discharge roll by cake-to-cake adhesion and is then scraped from the discharge roll by a doctor blade.

PRECOAT FILTER

Precoat filters are used when filtrate clarity is of prime importance, or in handling slurries where the solids content is less than 3% by weight and the cake is slimy or tacky. Precoat filters are also used when the formed cake is too thin to discharge continuously, or the filter cloth blinds after a short period of operation.

In precoat filtration, the filter drum is coated (before filtration is started) with a layer 1—6 in. (25—150 mm) of diatomaceous earth or other suitable material such as wood flour, potato starch or carbon (Figure 3.16). The choice of precoat material depends on the specific application. Care should also be taken to assure that the precoating liquor has the same characteristics as the process liquor. The precoat

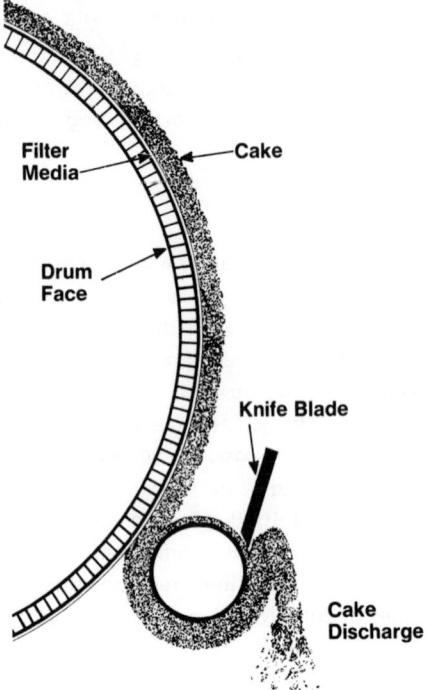

Figure 3.15 Roll discharge.

slurry is made up in a mix tank and pumped to the filter vat at a solids concentration of 2–10%. The suspension is filtered to develop a bed of precoat. Once the desired precoat thickness is achieved, the vat is drained of precoat slurry and the process stream is introduced to the vat. As the drum rotates, a knife blade positioned across the full width of the drum advances slowly, shaving off the filter cake and a very thin layer of the precoat bed, thus creating a fresh filtering surface each time the drum revolves.

The cut depth may vary from 0.0005–0.015 in. (0.0127–0.38 mm) and is set to cut slightly deeper than the actual penetration of filtered solids into the precoat bed. Mechanical stops prevent the blade from touching the drum, and thus leave a heel of precoat material at the end of the cycle. The life of a precoat cycle can be determined by the following equation:

$$T_c = \frac{P_c - h}{N_d C}$$

where T_c = cycle time, hr
       P_c = precoat depth, in.

h = heel thickness, in.
N_d = drum speed, rev/hr
C = precoat cut, in.

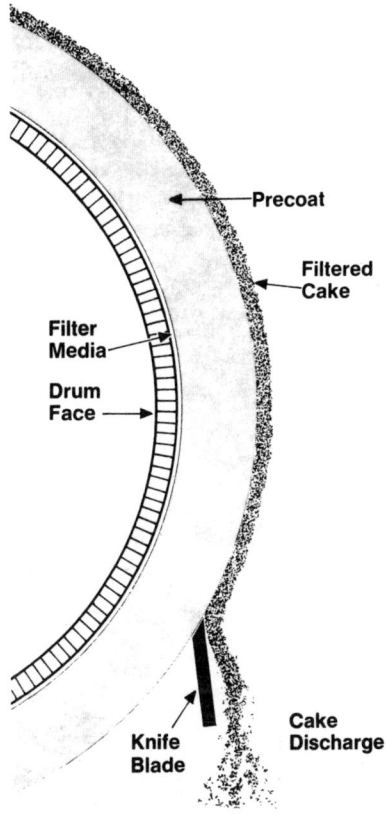

Figure 3.16 Precoat discharge.

REMOVABLE-MEDIUM FILTERS

The discharge mechanisms discussed thus far have the common characteristic of a filter medium permanently fixed to the drum. This permits the washing on only one side of the medium and may, in many instances, result in progressive blinding, which necessitates eventual replacement of the filter medium. This characteristic led to the development of drum filters on which the filter medium leaves the drum, is washed and is reapplied in each rotational cycle.

COIL-SPRING FILTER

A coil-spring filter (Figure 3.17) is a rotary drum vacuum filter on which the filtering medium consists of two layers of stainless steel helically coiled springs placed in corduroy fashion and around the drum as shown in Figure 3.18. The filter cake is formed on top ofof the coilsprings as the drum rotates through the vat. As the point of discharge is approached, the two layers of springs leave the drum deck. The upper and lower layers separate, and the upper layer carrying the cake passes over a discharge roll where the cake is removed. The lower layer passes around its own roll, is washed and returned to the drum over an alignment roll. The top layer of springs is also washed and returned to the drum over an aligning roll. (The discharge roll arrangement is shown in Figure 3.19).

Figure 3.17 Rotary drum vacuum filter with coil spring discharge. (Komline-Sanderson).

The coil-spring filter finds its best application in the dewatering of slurries that have a fibrous nature or that can be chemically flocced, such as the waste from a pulp mill or sewage treatment plant. This filter has the advantage of a filter medium that is essentially permanent and nonblinding. Yield rates are unaffected by grease, oils and other substances that might affect synthetic media.

CLOTH BELT DISCHARGE

The cloth belt discharge method (Figure 3.20) utilizes a filtering medium in the

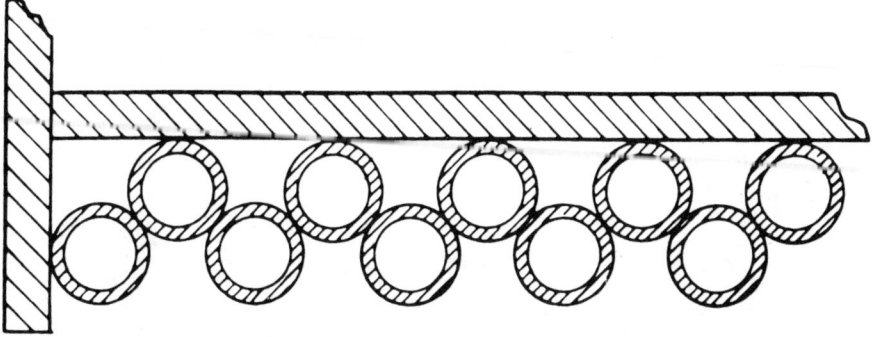

Figure 3.18 Cross section of coil spring filter medium laid corduroy fashion around filter drum.

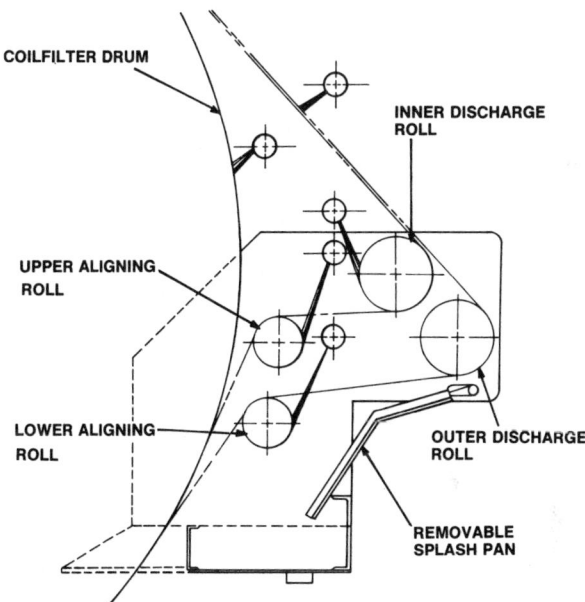

Figure 3.19 Schematic elevation of Coilfilter® (Komline-Sanderson).

form of a synthetic cloth belt equipped with a closure to make it endless. The filter cake is deposited on the belt surface, and the belt leaves the drum surface at the end of the "drying zone." It then passes around a discharge roll where the cake is removed during the approximate 120° change in the direction of belt travel. On leaving the discharge roll, the bet travels around a wash roll and then over a return

Chapter 3. Leonard P. Egee

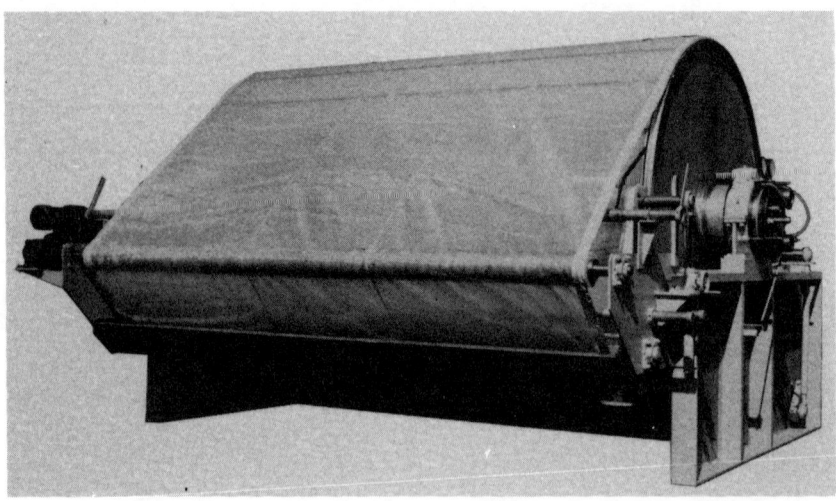

Figure 3.20. Rotary drum vacuum filter with cloth belt discharge. (Komline-Sanderson).

roll, which leads it on to the drum surface to repeat the cycle. As the belt travels from the wash roll to the return roll, it passes between spray headers and is washed on both sides.

SINGLE-CELL ROTARY DRUM FILTER

The single-cell filter (Figure 3.21) has no internal drum piping and does not use the conventional filter valve. Instead, the entire filter drum operates under vacuum and consists of a perforated cylinder having a large number of drainage panels on its outer surface. These panels are 2–2½ in. (50–60 mm) wide, or about one-tenth the width of conventional rotary drum filter compartments. The filter medium is supported on snap-in drainage plates made of punched metal and is secured to the surface by rods holding the fabric in the division strips.

Cake discharge is effected by air blowback from an internal "shoe" mounted inside the drum at the point of discharge. The blowback shoe, or valve, is fitted with very close clearances to the accurately machined inside surface of the drum, thus sealing off the vacuum at the point of cake discharge. The shoe has a narrow longitudinal slot across the entire drum width through which a large volume of low pressure air can be directed to the underside of the cake, forcibly removing it and cleaning the filter medium. Vacuum is applied to the inside of the drum through a large trunnion, and full vacuum is exerted directly under the filter medium without passing through piping or valves.

A rugged, stationary axial pipe supports the blowback shoe and also acts as a channel for filtrate and vapor. The open interior construction and the large center pipe allow this filter to handle large flowrates of both liquid and vapor.

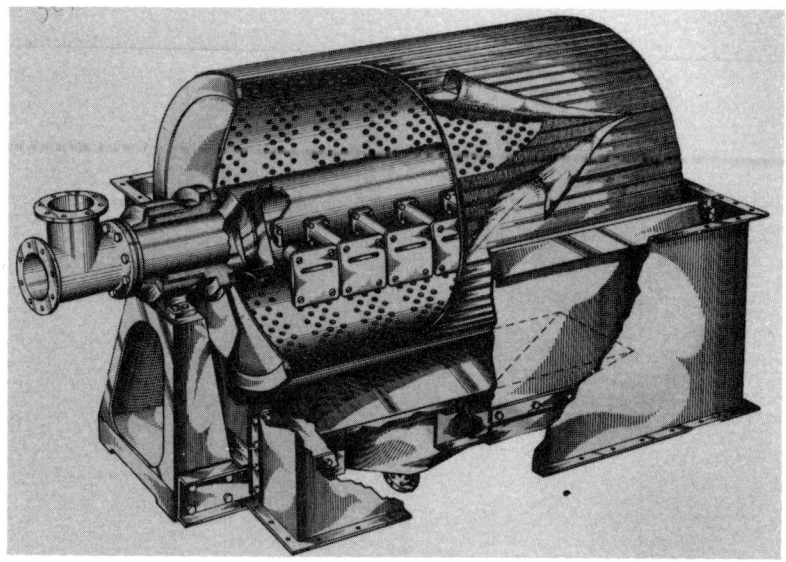

Figure 3.21 Bird-Young single-cell rotary drum vacuum filter. (Bird Machine Co.).

The inside of the drum acts as a receiver for separating air and filtrate, the latter being removed by an internal siphon extending through the shaft. This feature permits operation of the filter with thin cakes, so that high drum speeds (up to 26 rpm) and substantially greater capacities result. It can be employed on both slow and rapid-filtering materials, although it has limited flexibility due to the elimination of the valve and individual section control. It is available in either an open-type design or completely enclosed for vapor-tight or pressure operations up to 150 psi and temperatures of 177°C (350°F).

The "single cell" filter is available in sizes up to 140 ft^2 (13 m^2). Its chief advantages are:

1. high capacity per unit area;
2. efficient cake washing because of ability to handle thin cakes; and
3. low internal pressure drops

A type of unit without the internal shoe is also in use (especially in Europe) due to the lower cost. Without the air-blow discharge, it must be operated as a precoat filter using the product for precoating and a conventional scraper to continously peel off a thin layer of cake. Eventually (after 8–16 hours of operation), the product precoat becomes sufficiently plugged with fines to require renewing. The filter is stopped and, after the old precoat is dropped, a new precoat is applied.

TROMMEL® FILTER

The Trommel® filter (Figure 3.22) consists of a perforated drum suspended in an open tank, which contains the slurry to be filtered. The drum is covered with newsprint (or other type of filter paper) and rotates slowly through the slurry. The wetted paper seals against the ends of the drum (end seal strips) to prevent solids leakage into the filtrate.

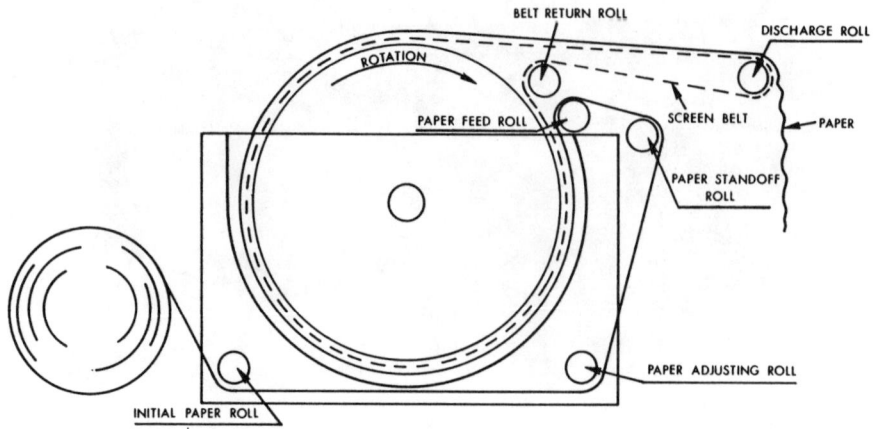

Figure 3.22 Trommel belt and paper tracking schematic. (Technical Fabricators).

Vacuum, which is applied to the drum internally, causes the liquid portion (filtrate) of the slurry to pass through the paper. Solids contained in the slurry are deposited on the paper.

The paper containing the dewatered solids or "cake" is discharged from the drum by a carrier belt. The carrier belt is returned to the drum and fresh paper is fed continuously to it. A seal plate is used in the area between the paper takeoff and the slurry level to prevent air leakage and vacuum loss through the exposed surface of the drum.

The filtrate is carried through the drum drainline spokes, hollow center shaft and rotary joint valve, to the filtrate receiver. At the receiver, the air and filtrate separate; the air being pumped through the vacuum pump and the liquid through the filtrate pump. If required, an oscillating rake-type agitator is provided in the filter tank to keep the slurry solids in suspension.

TOP-FEED FILTER

For coarse, fast-settling solids that dewater readily, the top-feed filter may be considered. This filter, with a drum similar to the standard rotary drum filter, does not have a slurry vat. Slurry is fed near the top of the drum and is confined between the drum end flanges and seals on the feed box. The cake is generally discharged into a chute located at the bottom of the drum.

For use with free-draining materials, it is not uncommon to modify the top-feed filter to a combination filter-dryer. In these applications, the filter is totally enclosed and hot air is introduced so that extremely low cake moistures (3% or less) may be obtained. This unit will handle the fast-settling materials that a standard drum filter cannot, and overcomes the particle degradation and dusting that occur with other methods of dewatering and drying crystalline materials.

HOPPER DEWATERER

The hopper dewaterer is a modification of the top-feed filter. Here, the drum surface is divided into hoppers that are 6 in. or more in depth. Slurry is fed to each hopper as it approaches the top of the drum. Generally, there is a brief gravity-drainage period after which vacuum is applied. The vacuum is released to allow gravity discharge as the hopper approaches the six o'clock position.

INTERNAL DRUMS

A vacuum filter of the rotary drum type, with its filtering surface arranged in panels on the inside of the drum, is known as the Dorrco filter. This filter has no slurry vat, as the material to be filtered is fed to the inside of the drum. One head of the drum is generally closed while the other is partly closed by an annular retaining ring through which the feed in introduced and cake is discharged.

The drum rests on external rolls, which are driven by an adjustable-speed drive. The filtrate piping is externally mounted at the closed end of the drum, and terminates in a conventional filter valve. Cake discharge is by a scraper and a chute. In the case of long drums, a belt conveyor may be used.

These filters are used for fast-settling materials and are low in initial cost, as they do not require a slurry vat or agitator. They are little used today because of the following disadvantages:

1. Cake washing is impractical.
2. Only a limited area of drum can be used.
3. Cake may fall off causing vacuum to drop.
4. Discharge can be difficult.
5. Replacing the filter medium is difficult.

ROTARY DISC VACUUM FILTER

The rotary disc vacuum filter (Figure 3.23) is composed of one or more filter leaves or discs mounted vertically on a horizontal shaft and suspended in a filter tank. Each disc is divided into a number of separate sectors that have suitable drainage and filter medium support. The sectors are held in place by tierods with the outlet nipple mating with a socket in the center shaft, which connects it to the filter valve. Sectors are readily removable and interchangeable. If a cloth filter medium is used, it is generally in the form of a bag that slips over each sector. Wire fabrics, when used, are welded or soldered to the sector.

Figure 3.23 Rotary vacuum disc filters operating in a Taconite plant. (Eimco Corp.).

Disc filters are available with discs from 4–12 ft (12–3.6 m) in diameter and with 1–20 discs providing filtration areas of 100 ft^2 (9.1 m^2) to 4,200 ft^2 ft. (380 m^2). Because the disc filter has drainage area on both sides of each sector, it offers more filtration area for a given floor space than the drum filter. This factor varies from 2.25–4.0, depending on the filter size. It is necessary to operate such units at high submergence, as a sector must be completely immersed during the cake-forming portion of the cycle. This reduces the amount of the cycle available for drying time and results in a ratio of drying time to form time of 1:1.

Cake discharge is effected by a scraper blade, although tapered roll discharges are sometimes used. An air blowback is often applied to assist the scraper or roll. In some instances, discharge is by high-pressure sprays. The discharge side of the vat is built to provide a discharge chute between the discs through which the cake falls to a conveyor below. Disc vacuum filters are customarily used for handling large volumes of free-filtering materials and have found most of their application in the metallurgical field.

HORIZONTAL VACUUM FILTERS

Horizontal vacuum filters, which utilize a flat, horizontal surface as the filtration area, can be divided into two major types, each with subdivisions:

1. Horizontal rotating pan
    a) Scroll discharge
    b) Tilting pan discharge
2. Horizontal "endless cloth"
    a) Traveling belt support
    b) Vacuum tray support

The horizontal vacuum filter was developed for those applications that could not be carried out practically on continuous rotary drum vacuum filters, and to replace filter presses to reduce labor costs. Rapidly settling materials, which cannot be filtered on rotary drum vacuum filters, or with great difficulty on plate-and-frame filter presses, are handled very successfully on horizontal vacuum filters. When the slurry is fed to the filter, the heavy fractions tend to settle, quickly acting as a precoat. On top of this bed, the medium and fine fractions settle, yielding ideal stratification. Fragile crystals will often give a better filtration rate on a horizontal vacuum filter than on a rotary drum vacuum filter.

In many continuous filtration applications, cake washing is of primary importance. On a horizontal vacuum filter, high cake-washing efficiency is possible because the cake can be flooded with wash liquor, which allows for plug flow. Sharp separation of filtrates is possible, permitting several stages of cocurrent or countercurrent washing. It has been demonstrated that in certain processes with critical washing requirements, one horizontal vacuum filter can replace two or more rotary drum vacuum filters.

Horizontal vacuum filters usually cost more per unit of filtration area and require more floor space than rotary drum vacuum filters. As indicated in the preceding paragraph, the total cost of an installation of horizontal vacuum filters may be less than that for rotary drum vacuum filters operating on the same process.

HORIZONTAL ROTATING PAN

This category of filters is typified by a circular, horizontal pan, which rotates around a hub containing the filter valve. The filter medium is fixed to the wedge-shaped sections of the pan, feed is applied to the top of the cloth and vacuum is applied from below. The major distinction between machines in this group is by method of discharge.

Horizontal Rotating Pan with Scroll Discharge

The horizontal pan scroll discharge filter (Figure 3.24) is a rotating horizontal

Figure 3.24. Horizontal rotating pan filter with scroll discharge.

"table" divided into sectors by division strips with each sector being a separate compartment. Vacuum is applied to each compartment through a filter valve located under the center of the table. The bottom of each compartment slopes toward the center of the filter and drains into the valve. The filter medium which can be either a wire screen or a cloth, is supported by a drainage grid and is caulked into dove-tailed division strips.

The slurry to be filtered is applied by a pipe and weir box distributor above the filter. The filter cake is discharged by a spiral scroll just ahead of the feed zone. A dam is placed between the scroll and the feed area to prevent the slurry from running back to the cake discharge zone. Control of this type of filter is generally effected by adjustment of rotational speed with a variable-speed drive and by rate of slurry feed.

Active filtration areas range from 4 ft^2 (0.37 m^2) to 500 ft^2 (46.5 m^2) with some designs ranging up to 100 ft^2 (92 m^2). This type of filter is usually limited to slurries that can form a ¾-in. (19 mm)-thick cake in 30—60 seconds. Various materials of construction are available, and high vacuum levels are possible.

A major problem associated with the scroll discharge is the fact that the scroll does not completely discharge the cake. There must be a clearance of approximately 1/8—1/4 in. (3—6 mm) between the scroll and the filter medium, which means that a layer of cake must remain on the cloth. This layer, or heel, may tend to smear or compress, due to the wiping action of the scroll, and adversely affect the filtration rate. The heel is carried into the feed zone (which means the cloth is not washed), where a reverse air-blow tends to dislodge and remix it with the incoming feed. This feature is used to minimize cloth blinding and maintain capacity.

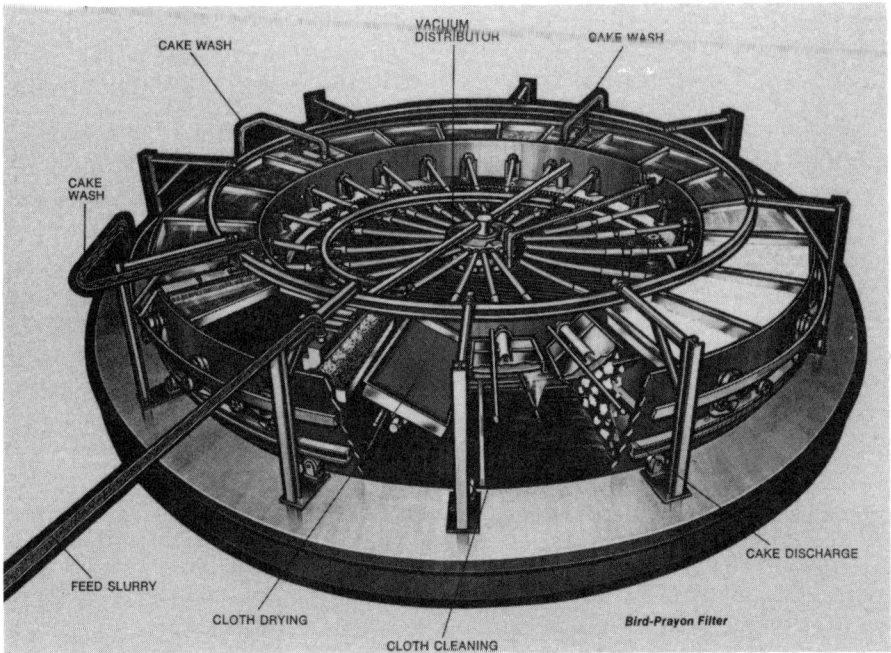

Figure 3.25 Bird-Prayon horizontal vacuum filter with tilting pan discharge. (Bird Machine Co.).

Horizontal Rotating Filter with Tilting Pan Discharge

The tilting pan horizontal filter (Figure 3.25) is similar to the horizontal scroll discharge unit, in that the filtration surface is a circular horizontal surface divided into segments. The essential difference between these two units is that the tilting pan filter compartments are independent pans connected to a center valve by a swivel pipe joint, permitting them to tilt around a radial line through their centers, whereas the scroll discharge type has one circular pan divided into compartments by division strips. Cake formation and washing take place in a manner similar to that on the scroll discharge machine, but the cake is removed by inverting each pan as it passes the point of discharge. An air blowback is sometimes employed to assist with cake release, and cloth washing on one side is possible to prevent blinding.

This filter is available with active filtration areas from 200–2,730 ft^2 (18–254 m^2) with a variety of materials of construction. It can operate with vacuums up to 24 in. Hg. This filter is used with slurries which will form a cake at least ½ in. (12 mm) thick in about 30 seconds, and has been applied primarily in phosphoric acid filtration (filtering gypsum).

Compared with a scroll discharge filter, the tilting pan is more expensive per unit area, will require approximately twice the floor space, and has relatively high maintenance costs. It does allow, however, for excellent control of wash liquor, produces excellent cake release and is available in much larger sizes. Field assembly is possible.

HORIZONTAL ENDLESS CLOTH VACUUM FILTERS

Horizontal endless-cloth vacuum filters are well suited to coarse, freely draining materials or fine, slowly draining solids, especially where washing requirements are critical. Sharp separation of filtrates can be obtained so that multistage countercurrent washing can be carried out effectively.

Whereas the types of rotating pan horizontal vacuum filters can be subdivided by method of discharge, the endless-cloth filters are classified by the method employed to support the filter medium. In essence, they resemble a conveyor with the feed entering at one end, mother liquor removed, wash liquor applied (as necessary) and dry cake being discharged by gravity as the medium passes over the discharge roll. This type of filter requires approximately the same amount of floor space as the tiliting pan filter and costs substantially less.

Filter Medium Supported by a Traveling Belt

This category of filters is typified by an elastomeric belt under tension (300 lb/in. of width) between two large-diameter rolls and supporting the filter cloth.

Feed is distributed by means of a weir. Side dams are necessary to contain the slurry and wash liquor until a dry cake has been formed. Cake is discharged as the cloth passes over a small diameter end roll. As the cloth returns to the feed end, it is spray-washed on both sides to pevent blinding. Filter sizes range from 5–2,150 ft^2 (0.5–200 m^2). Hooded enclosures are available.

The rubber belt, although reasonably durable, is sometimes a point of concern with this filter. The material of construction limits the industrial applications where solvents may be present, and maintenance tends to be high. Because the rubber belt is constantly moving, there is a continual vacuum leakage between the belt and the stationary vacuum box, which distributes the vacuum and separates the filtrates from multistage washing. This leakage increases the vacuum capacity required, and generally limits the vacuum level to 20 in. Hg.

Filter Medium Supported by a Vacuum Tray

This grouping can be broken into two subgroups: one in which the trays reciprocate, and one in which the belt indexes. Both are continuous filters, even though vacuum is intermittent. Because there is no relative motion between the cloth and vacuum trays while the vacuum is appliied, both machines are able to maintain very high vacuum levels. Since there is "dead time" (when the vacuum is vented), this portion of the cycle must be taken into account when sizing a unit.

The vacuum trays are fabricated with integral side dams so that slurry is positively contained. Washing can be performed effectively and, since the rubber belt has been eliminated, exposure to organic solvents is permissable.

Modular construction not only allows for field assembly, but future expansion, if process requirements change.

Reciprocating vacuum tray — The reciprocating vacuum tray type of filter uses an endless filter cloth, which is supported by the vacuum trays rather than by a rubber belt. In this system, the vacuum trays move forward with the cloth while the vacuum is applied and return rapidly to their original position after vacuum has released, following which the cycle is repeated.

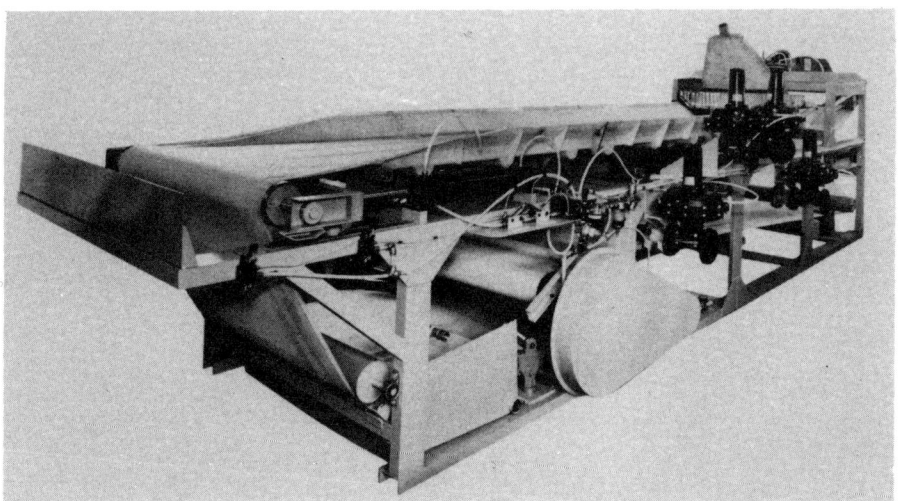

Figure 3.26 Horizontal endless cloth vacuum filter with medium supported by vacuum tray. (Komline-Sanderson).

The development of the reciprocating vacuum tray was precipitated by the need to do away with the rubber support belt in solvent-washing applications. Mechanically, this filter is more complex than the rubber belt type and, therefore, more expensive per unit area. It is limited to 80 in. (2 m) widths and lengths of approximately 64 ft (20 m).

Cake is discharged over a 20-in. (0.5 m)-diameter roll, which means that thin cakes may be troublesome. Efficient cake washing can be achieved by any of a variety of arrangements. Two-sided cloth washing can be accomplished while the cloth travels from the discharge roll to the feed end.

Indexing cloth — The next stage in the development of endless cloth type horiziontal vacuum filters was the indexing cloth machine (Figure 3.26). Stationary vacuum trays support the cloth, which is indexed by means of a reciprocating discharge roll. During the period in which the vacuum is applied to the slurry, the cloth is stationary on vacuum trays, the discharge roll is slowly retracting and the cloth moves under the machine back to the feed area. The vacuum is then vented, the discharge roll rapidly advances, indexing the cloth, and the cycle then repeats. Two-sided cloth washing is possible in the same manner as on the reciprocating vacuum tray type. The major advantages of this filter are its relatively simple design and low maintenance costs.

CHAPTER 4

CENTRIFUGES

W. F. WHITE
Bird Machine Company, Inc.
So. Wapole, MA

INTRODUCTION

Centrifuges have been highly developed for the use of liquid-solid and liquid-liquid separations in the chemical process industries. A centrifuge is a piece of equipment that employs centrifugal force to separate the desired materials. It is found in a wide variety of types and sizes; the problem for the process engineer is to select the type most suitable for a specific separation problem.

The centrifugal forces employed in these equipments vary from less than 100 X g up to 10,000 X g, or times the gravitational acceleration. The centrifugal force is proportional to the square of the speed, and radial distance from the center of rotation. It may be calculated as follows:

$$F_c = (rpm^2) \times (D \text{ of bowl in mm}) \times (5.47 \times 10^{-7})$$
$$= \frac{(rpm^2) \times (D \text{ of bowl in ft})}{6000}$$

Liquid-solid separation centrifuges can be broadly broken down into two types: (1) sedimentation centrifuges, which depend on the sedimentation of solids against a centrifuge bowl wall; and (2) filtering centrifuges, in which the liquor being separated must filter through both the solids being separated and a filter media on the centrifuge wall. Liquid-liquid centrifuges and liquid-liquid-solid centrifuges are of the sedimentation type. Table 4.1 shows the major types of centrifuges employed in the chemical process industries and illustrates the type of separations in which they are normally employed (together with the usual capacities), a description of their general method of operation, as well as their relative efficiency on washing and the effect on particle degradation of material being processed. This table may be employed as a preliminary screen in the selection of suitable centrifuges.

FILTERING CENTRIFUGES

Filtering centrifuges are normally employed on the separation of coarse particles greater than 44 μ (325 mesh). Their solids-handling capacity is greatly affected by

Chapter 4. W. F. White

Table 1.

Type	Usual Maximum Centrifugal Force Level 1000 x G	Particle Size Range	Usual Concentration of Feed Solids %	Capacity M³/Hr. (GPM)	Capacity Ton/Hr Dry Solids Discharge	Relative Level of Effluent Clarity	Solids Discharge Type	Solids Discharge Character	Relative Efficiency of Washing	Particle Degradation	Equipment Manufacturers
Pusher	1000	<75µ	30-60	—	½-60	Low	Continuous	Solid	High	Low	Baker-Perkins Bird, DeLaval
Scroll-Screen	1000	<150µ	>50	—	½-40	Low	Continuous	Solid	Low	Medium	Dorr-Oliver, Heinkel, Pennwalt
Pealer-High Speed Unloading Basket	1000	<75µ	20-50	—	½-10	Medium	Controlled Cycle	Solid	High	High	Baker-Perkins Pennwalt
Basket-Low Speed Unloading	1000	<44µ	10-50	—	.1-1.5	High	Controlled Cycle	Solid	High	Minimum	Ametek, DeLaval Pennwalt
Oscillating/Tumbler	300	<300µ	50-75	—	30-100	Low	Continuous	Solid	Low	Low	Ametek, Bird Envirotech
Screen Bowl	2000	<44µ	2-40	¼-200 (1-800)	½-100	High	Continuous	Solid	Medium	Medium	Bird, Pennwalt
Solid Bowl	3000	<1µ	2-40	¼-250 (1-1000)	¼-60	High	Continuous	Solid	Low	Medium	Bird, DeLaval Dorr-Oliver
Disc-Nozzle	5000	75 to .1µ	2-10	¼-180 (1-800)	—	High	Continuous	Fluid	Low	—	Centrico, DeLaval Pennwalt
Disc-Self Opening	5000	75 to .1µ	2-10	¼-50 (1-200)	—	High	Controlled Cycle	Semi-Fluid	Low	—	Bird, Centrico DeLaval
Tubular	10000	300 to .1µ	0-.05	¼- 5 (1-20)	—	High	Manual Removal	Solid	—	—	Delaval, Pennwalt
Vortex Clarifier	3000	100 to .1µ	>.01	2-250 (10-1000)	—	High	Controlled Cycle	Fluid	—	—	Bird

the solids concentration being fed to the equipment. The higher the solids concentration, the higher the capacity. In some instances a 20% reduction in solids concentration will result in a 50% reduction in solids-handling capacity of the equipment. In addition, the losses through the screens are also normally affected by the solids concentration, so that the higher the solids concentration, the less the losses are through the screens or filter medium on the centrifuge. In most instances, filtering centrifuges should be considered part of a system that would include a settling device or hydrocyclone, to prethicken the feed to the centrifuge. Some typical examples are shown in Figure 4.1. The effluent of the centrifuges in these systems is returned to the feed concentrator.

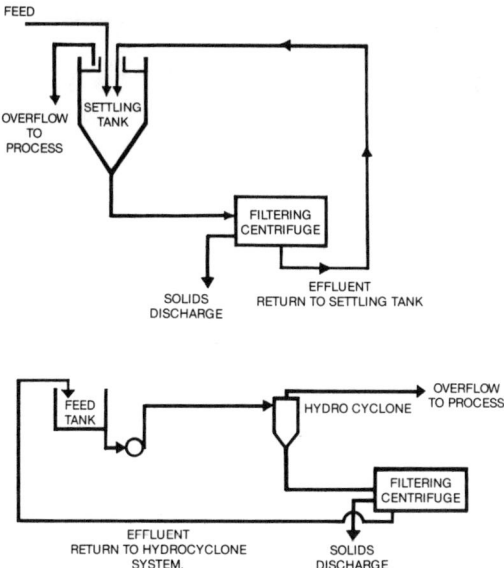

Figure 4.1. Typical filtering-type centrifuge system.

Pusher Centrifuges

The pusher centrifuge variety of filtering centrifuge is so-called because of the pushing mechanism employed to transport the solids across the basket. Pusher centrifuges are normally employed when the feed can be concentrated above 60% by volume, as their capacity is greatly increased as the concentration is increased. They have a relatively high capacity on concentrated feeds and provide for an efficient washing of the solids where required. They are also capable of handling fragile crystals, which may be broken or smeared in other types of centrifuges. Typical of these are caffeine, crude sodium bicarbonate and other long-needle crystals.

In some unusual circumstances, it may be desirable to preconcentrate the feed to the pusher type centrifuge through a vacuum filter. Such a system is shown in Figure 4.2. This will provide a very high degree of clarity of the effluent discharged from the system. It can be used for efficient, multistage washing and can provide very high capacities.

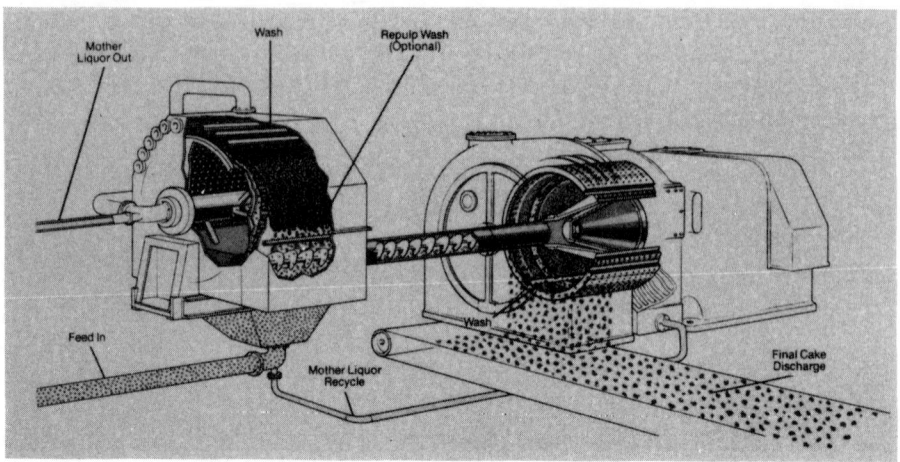

Figure 4.2. System of vacuum filter and push-type centrifuge. (Photo courtesy of Bird Machine Company, Inc.)

A two-stage, push-type centrifuge is schematically shown in Figure 4.3. The operation of this equipment is as follows. Feed enters through the inlet pipe (1) and accelerating cone (2), and is introduced on the first stage basket (3) where the solids are retained on the sieves. The first stage basket, actuated by the hydraulic pushing mechanism (4) reciprocates under a static pusher plate (5) to advance the cake from the first to the second stage on the back stroke. The forward stroke of the first basket pushes the cake off the second basket and into the cake chute (6). In models with more than two stages, alternate baskets telescope, easing the cake from stage to stage. In multistage units, multiple washes can be employed (7), and mother and wash liquors can be separately discharged, if desired, through a compartmented effluent chamber (8). A single-stage, push-type centrifuge is shown in Figure 4.4.

Push-type centrifiges are manufactured from 200–1,200 mm (8–48 in.) in diameter. The smallest centrifuges would handle approximately 200 lb/hr, the largest up to 60 ton/hr, under proper conditions. Capacity and performance will vary widely with the particle size and shape of the material being separated; the coarser the material the higher the capacity and the lower the moisture.

Push type centrifuges are built with either a single stage, that is, only one basket,

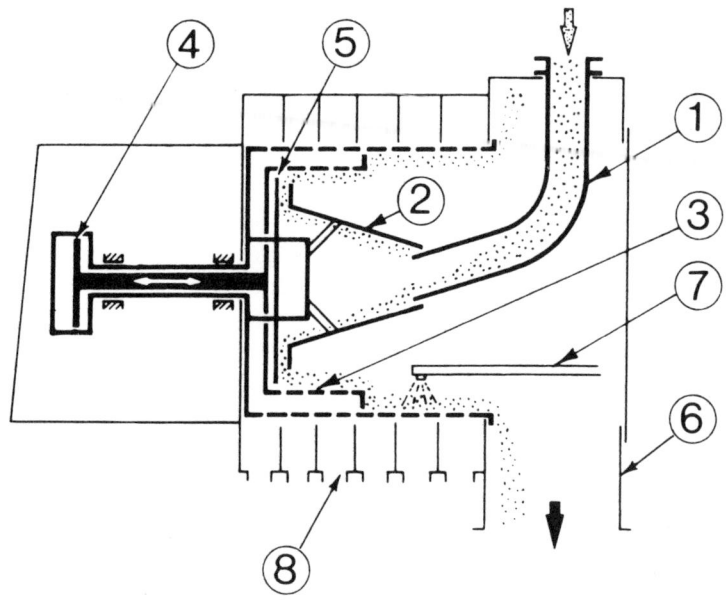

Figure 4.3 "Two-stage push-type centrifuge — how it works."

Figure 4.4 Single-stage, push-type centrifuge. (Photo courtesy of Baker-Perkins Inc.).

or multi-stage with two or more baskets. The purpose of the multi-stage design is to break down the basket length. The cake thickness on the basket is generally proportional to the friction of the cake over the filter sieves, the centrifugal force applied, and the length of the basket. In large single-stage centrifuges, the load in the basket to be pushed is greater than can be pushed by the pusher plate through the relatively plastic cake in the rear of the basket. This may cause cake buckling which can result in channeling of the feed slurry through the centrifuge cake, and subsequent imbalance. These unbalances on this cantilever design centrifuge causes severe vibration which usually requires shut down and wash out of the centrifuge.

Scroll Screen Centrifuge

The scroll screen type centrifuge consists of a conical basket that can be either horizontal or vertical and contoured to fit inside the cone is a scroll. Depending upon the product handled and the angle of the cone, the scroll can be used either to retard the advance of the solids through the basket, or to transport the solids down the basket. They are not as susceptible to feed concentration as the push-type centrifuge, although solids losses and capacity will be greatly affected by feed concentration. Relatively high capacities can be obtained at a moderate price in this type of equipment. The effectiveness of washing is relatively low, both because of the short retention time in the machine, and the relatively poor distribution of wash over the surface of the product being transported. When fragile crystals are handled the action of the scroll across the surface of the screen tends to break the particles and drive them through the screen. Where fine separations are employed (below 300 μ (50 mesh) relatively fine screens are employed, and if the product being handled is abrasive, short screen life can often be anticipated.

A scroll screen centrifuge is shown in Figure 4.5. The feed comes into the feed pipe at the top of the cone at the small diameter where the bulk of the mother liquor is removed as the solids accelerate down the screen. The scroll is run at a speed either slower or faster than that of the bowl, the differential movement generally being supplied by a cyclo gear, which provides the differential movements between the bowl and the scroll. The sizes offered vary from 250 mm (10 in.) in diameter with a capacity over 3,000 lb/hr up to large machines employed typically in the coal industry of 900 mm (36 in.), which handle 20—30 ton/hr.

Another type of screen centrifuge is illustrated in Figure 4.6. This type of centrifuge does not employ a scroll but has a series of baffles that rotate at the same speed as the basket to slow the path of the solids as they progress along the basket and over the screen. This provides the desired retention time so that the desired dewatering can be accomplished. The retention time is regulated by adjusting the baffles. Performance is similar to other scroll screen machines and eliminates the possible degradation of particles caught between the scroll operating at a different speed than the basket.

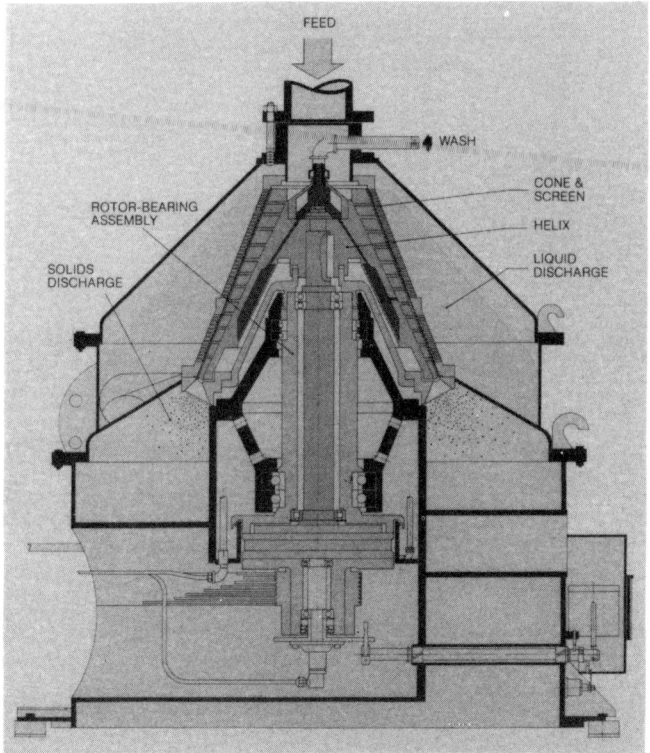

*Figure 4.5 Scroll screen centrifuge.
(Illustration courtesy of Dorr-Oliver Inc.).*

Basket Centrifuge

The basket centrifuge represents the original type centrifuges used for liquid-solid separations. They are used extensively throughout the process industries worldwide. Most of the centrifugals for sugar are of this type. They are manufactured in an "overdriven" type with the motor and shaft above the basket, and the basket suspended from an overhead frame; or an underslung type, in which the basket is driven from below, usually suspended with a three-point suspension from the frame.

In recent years, except for sugar centrifugals where the highest forces are employed, the underdriven machines generally have been used. They have been developed so that they are available with completely automated systems, so they can be fully programmed to go through the desired cycles without operator attention. They are constructed with both perforate baskets, in which a filter medium is installed. The liquor filters through the basket or imperforate versions where the solids are sedimented against the bowl wall and the effluent overflows the basket

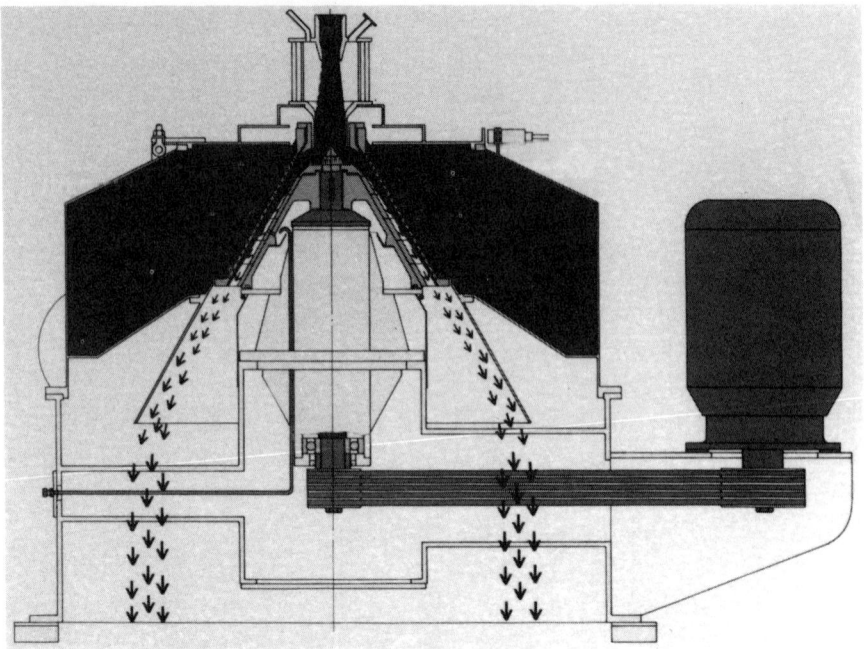

Figure 4.6 Screening centrifuge. (Photo courtesy of Ernst Heinkel Machinebau KG.)

top ring. Generally, the machines employed in the chemical and pharmaceutical industry are the perforate type used for separation of crystalline or other discrete particles, whereas the imperforate type is generally empolyed in waste treatment application and for the dewatering of very fine particles and/or sludges. These machines are batch types, and the cycles are almost infinitely variable. The cycle can be varied to achieve the desired performance. The basket provides a good surface for washing the material contained therein. Long spin times can provide maximum dryness.

The baskets are generally built from 12–48 in. in diameter, the largest generally employed. These machines will usually be able to handle a capacity of about 12 ft^3 of product per cycle. The electrical or hydraulic systems on the underdriven machines generally have a minimum cycle of about 12–15 minutes. Unloading is accomplished at a speed of about 50 rpm. Overdriven machines with very large motors can be designed for shorter cycles.

This type equipment is widely employed in the fine chemical and pharmaceutical industry, where products are produced on a batch basis and many different products are produced with the same equipment. The variable feature of the cycle enables one to change the cycle from batch to batch or product to product as required.

Peeler Centrifuge

The peeler centrifuge is another type of basket centrifuge. The basket usually rotates on a horizontal axis. The difference between this equipment and the previously described basket is that on the peeler the basket unloads the cake at full speed. A peeler centrifuge is shown in Figure 4.7. The knife moves into the basket to unload the solids at the full operating speed of between 800–1,200 rpm. This provides the capability of short cycles, usually in the range of 2–5 minutes, and provides for higher capacities than low-speed unloading baskets. Longer cycles can be employed; however, generally if these exceed 10 minutes, an underdriven machine is often the more economical choice.

This type equipment is still fairly extensively used in starch processing. The peeler provides close to continuous operation at relatively high capacities. In recent years many of the applications where this equipment was formerly employed are now done on continuous centrifuges.

Figure 4.7 Peeler centrifuge. (Photo courtesy of Baker-Perkins Inc.).

Miscellaneous-Type Centrifuges

There are several types of centrifuges that employ a conical basket. One of these is commonly known as a "skid pan," which consists of a conical basket rotating on a shaft fed at the small end of the cone. The liquid passes through a screen and the particles slide up the side of the basket where they are discharged. These were developed in the sugar industry where the feed conditions are highly controlled and consistent. The travel of the particles up the side of the basket depends on the feed consistency and the size and shape of the particles. In attempts to employ these in the chemical industry, where the solids and feed consistency are variable, the use of this equipment has had very limited success.

Another centrifuge of this type is illustrated in Figure 4.8, which is an oscillating centrifuge. This is similar to the skid pan in that it has a conical basket, but in addition, the basket is oscillated at approximately 2,000 cycles per minute, which helps transport the solids up the basket. This type of centrifuge is widely employed in the coal industry and is used extenisvely for dewatering of 6 X 0.5 mm (¼ in. X 28 mesh) coal. It is built in basket sizes from 1,000–1,250 mm (40–48 in.). It has a capacity on fine coal (6 X 0.5 mm) of 80–150 ton/hr and requires a well-pre-thickened feed, usually 75–80% solids, and is only effectively employed on materials in the plus 0.3 mm (48 mesh) range.

Figure 4.8 Oscillating centrifuge. (Photo courtesy of Bird Machine Company, Inc.).

Some of these equipments have been employed in the potash industry for separation of coarse potash and coarse salt. They also have been used in the production of sea salt.

Another centrifuge of a similar type, again having a conical basket, is a tumbler in which the basket rotates in a manner designed to transport the solids up the basket. These have been employed mainly in Europe.

SEDIMENTATION CENTRIFUGES

Sedimentation centrifuges, as the name implies, are settling devices employing centrifugal force for the acceleration of the settling effect. All sedimentation centrifuges require the solids to be heavier than the liquid to effect the separation. There are a variety of types employing different means of removing the solids from the centrifuges. These many types are required for the great variety of separation problems that exist in the process industries. They are not usually sensitive in their solids handling capacity to feed concentration because the liquid does not have to filter through the solids or a filter medium. They are employed for a wide variety of separations from relatively coarse particles of 6 mm (¼ in.) down to submicron sizes. Flocculants are often employed to assist in the agglomeration of the particles to aid in the settling for very fine materials and many waste sludges.

Solid Bowl Centrifuges

One of the most versatile of continuous centrifuges is the Solid Bowl Centrifugal. This was developed in the 1930s and since has been employed extensively in the process industries.

Figure 4.9 shows a cutaway of this machine, and the following is a description of how it operates. The two principal elements of the Solid Bowl Centrifugal are the rotating bowl, which is the settling vessel, and the conveyor, which discharges the settled solids. The bowl has adjustable overflow weirs at its larger end for discharge of clarified effluent and solids discharge ports on the opposite end for discharging dewatered solids. As the bowl rotates, centrifugal force causes the slurry to form an annular pool, the depth of which is determined by the adjustment of the effluent weirs. A portion of the bowl is of reduced diameter so that it is not submerged in the pool and thus forms a drainage deck for dewatering the solids as they are conveyed across it.

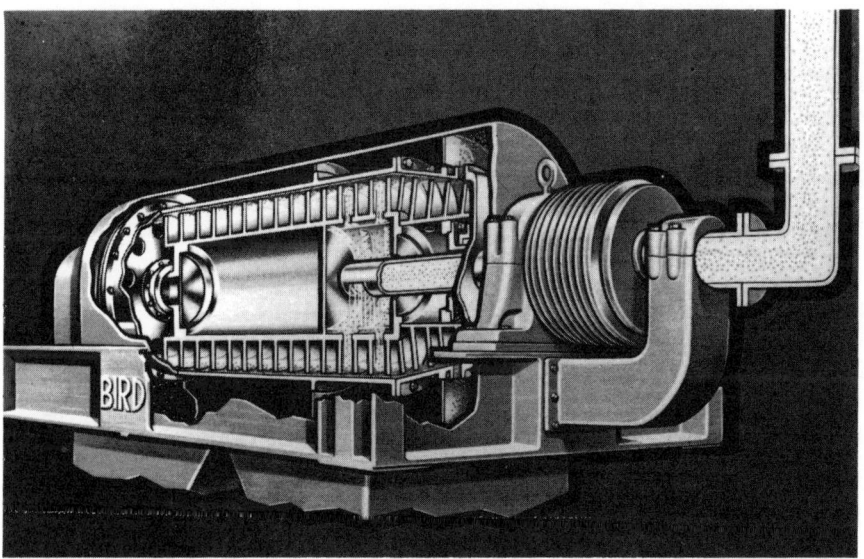

Figure 4.9 Solid bowl centrifugal. (Photo courtesy of Bird Machine Company, Inc.).

Feed enters through a stationary supply pipe and passes through the conveyor hub into the bowl itself. As the solids settle out in the bowl, due to centrifugal force, they are picked up by the conveyor scroll and carried along continuously to the solids outlet. At the same time, effluent continuously overflows the effluent weirs. If washing is required, wash liquors enter the centrifuge through a separate pipe and are sprayed onto the solids as they pass across the drainage deck.

Solid Bowl Centrifuges are made in sizes from 150–1,350 mm (6–54 in.) in diameter. They are made in lengths from 300–3,500 (12–140 in.) in length. They generally operate at speeds sufficient to generate centrifugal forces that range from several hundred up to 3,000 g's. They handle capacities from severl hundred kilograms (lb) to 100 ton/hr, and can handle hydraulic flows from several l/min (gal) up to 250 m^3/hr (1,000 gpm).

These machines have been developed with many modifications to enhance their performance. One special type produced has a concurrent flow through the machine. This is often employed on waste sludges and fine materials where flocculants are generally used, and can be advantageous in reducing flocculant cost and producing a better dewatered sludge.

Screen Bowl Centrifuge

Another version of this machine is called the Screen Bowl, and this is illustrated in Figure 4.10. The construction is generally similar to that of the Solid Bowl, except that after the solids have been removed from the pool, a screen section is employed where further washing and dewatering can take place. This, in effect, makes this a combination sedimentation and filtering centrifuge. It provides the advantage of both types of equipment. As mentioned under Filtering Centrifuges, the performance generally increases as the solids concentration is increased. In the Screen Bowl Centrifuge the discharge of a solid bowl is delivered to the screen portion. This provides for maximum dewatering with minimum retention time on the screen, it also minimizes solids losses through the screen. The benefits of additional dewatering and improved washing are achieved in the screen section of the centrifuge.

Disc Centrifuges

Another style of the basic sedimentation centrifuge is the high-speed, disc centrifuge. These were originally designed as cream separators, but through the years have been developed to a high degree of sophistication and have found many applications in the process industries. They are used for liquid-liquid separations, solid-liquid separations and for liquid-liquid-solid separations.

Generally speaking, disc-type centrifuges are used to separate and concentrate solids from slurries where the particle size is finer than 200 mesh down to submicron material. Only a very small-density difference is required to separate solids

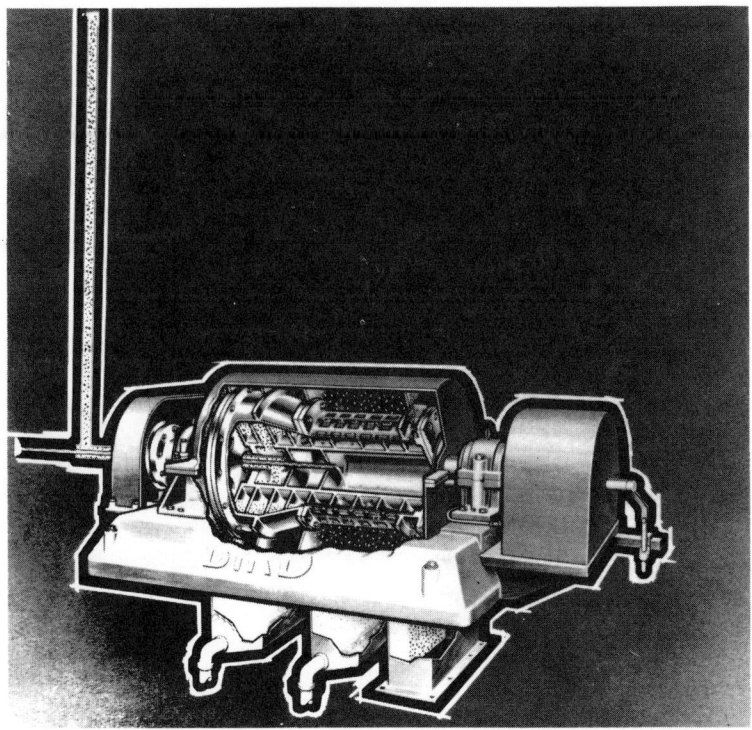

*Figure 4.10 Screen bowl centrifuge.
(Photo courtesy of Bird Machine Company, Inc.).*

in these high g machines. If oversized material is present, screens or other devices are used to remove oversized materials that might plug discs or nozzles in the machine. These centrifuges are generally built in sizes ranging from 200–1,000 mm (8–40 in.) in diameter. They have capacities varying from 200 l/min (0.1 gpm) up to 230 m^3/hr (1,000 gpm).

The disc-type centrifuge is built in several different types of bowls and discharge arrangements. The conventional liquid-liquid separator has no automatic sludge discharging means and must be dissassembled for sludge removal; thus, it is employed only where miniscule amounts of solid phase material are anticipated.

The "self-opening"- or "desludging"-type nozzle centrifuges are built with a reservoir to hold solids. Periodically, the bowl halves split to discharge the accumulated solids. This discharge can be controlled by a timer or by other controls triggered by the sludge volume accumulation in the bowl, or clarity of the overflow. These machines can store up to 30 liters of sludge and are used on slurries in the range of 1–10% solids by volume at feed rates up to 50 m^3/hr (220 gpm). A typical desludging-type centrifuge is shown in Figure 4.11. The arrangement illustrated is with a turbidimeter on the effluent to trigger the desludging cycle.

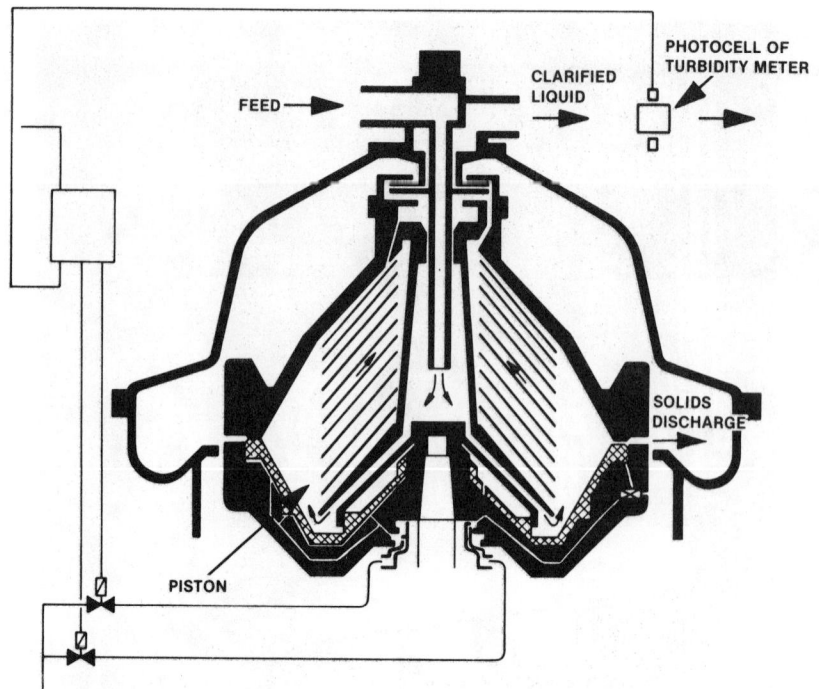

Figure 4.11 Desludging centrifuge. (Diagram courtesy of Centrico Inc.).

Referring to Figure 4.11, the feed is delivered into the spinning bowl where the liquid to be processed enters a set of closely spaced conical discs; here it is distributed into thin layers enhancing the action of centrifugal force. The solids are forced against the underside of each disc and slide into the sediment-holding space of the bowl. When sufficient solids have accumulated, a hydraulically operated piston is activated to open ports in the periphery, and the solids are rejected through the ports by centrifugal force. Opening and closing of the bowl are accomplished while it continues to run at full speed. An installation photo of a desludging centrifuge is in Figure 4.12.

The nozzle centrifuge is another type of disc centrifuge, and this is shown in Figure 4.13. Referring to this diagram, the feed enters the feed well of the high-speed rotor through a central passage. From the feed well, slurry enters the feed impeller, where it is brought up to rotor speed. Slurry then enters the separation chamber where centrifugal forces, thousands of times higher than gravity, cause the bulk of the solids to settle to the periphery of the rotor where they are continuously expelled through fixed open nozzles in the rotor to the underflow volute. Small quantities of lighter solid material, entrained in relatively clear liquid, are forced inwardly up the disc stack, when even the smallest particles become im-

Process Equipment Series Volume 1

Figure 4.12 An installation photo of a large desludging centrifuge. (Photo courtesy of Centrico Inc.).

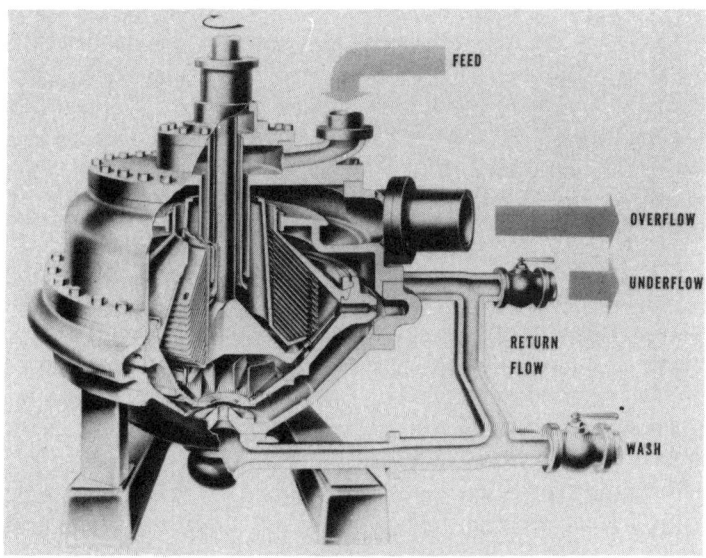

Figure 4.13 A nozzle-type disc centrifuge. (Photo courtesy of Dorr-Oliver, Inc.).

pinged on the underside of the discs. As the solids agglomerate and gain density, they fall, counter to the liquid flow, into the separation chamber to join the heavier material and be passed through the rotor nozzles. Thoroughly clarified, clear liquid continues up through the disc stack and out of the rotor into the overflow volute.

The nozzle-type centrifuge generally has a higher sludge-handling capacity than the self-opening-type machine. However, the self-opening- or desludging-type machine will produce a sludge of a higher solids concentration than that which can be obtained in the nozzle type. The usual operating speeds of these machines produce centrifugal forces in the range of 5,000—8,000 g's, depending upon the machine size.

TUBULAR CENTRIFUGES

Tubular Centrifuges are normally employed for either liquid-liquid separations or for removing small amounts of solids from liquids. They are, as described, a tube or cylinder with a retained volume. They are normally built in bowl diameters from 1¾—5 in. Some of the smaller ones develop in excess of 60,000 times g.

A typical Tubular Centrifuge is shown in Figure 4.14. The usual arrangement is to have the bowl suspended from an upper bearing through a flexible spindle support system. It attempts to find its natural axis of rotation if a slight unbalance occurs. Tubular Centrifuges are used in oil purification and in the pharmaceutical, food, paint and biochemical industries.

The feed is generally fed into the bowl continuously and the solids are sedimented out against the bowl wall while the liquid overflows the ports at the opposite end of the bowl from the feed. In some instances, a centripetal pump system is employed at the top of the bowl for removal of the discharged effluent.

When the bowl is full, the solids must be removed by disassembling the centrifuge and manually removing the plug of solids from the bowl. The capacity of these centrifuges varies from 1—80 l/min (¼—20 gpm). The solids handling capacity is quite limited, usually less than 2,000 g (5 lb).

A recent development, which is a type of a Tubular Centrifuge has been the Vortex Clarifier (Figure 4.15). It is a relatively large-diameter Tubular Centrifuge which employs stilling vanes to assure a forced vortex and to reduce turbulence. The feed enters the top through an impeller into accelerating vanes, which accelerate the slurry and feed it into the tubular section containing the stilling vanes. As in conventional Tubular Centrifuges, the solids are sedimented against the bowl wall. The liquid, on discharging, passes through power recovery vanes where virtually all the power expended in accelerating the liquid is recovered, and then the clarified liquid is removed through a volute. The typical cycles run from 30 minutes to many hours, depending on the solids concentration in the feed. The centrifuge is equipped with a timer cycle to program all of the operations of the equipment. When the cycle is complete, the feed is stopped to the machine, a brake is applied and the shaft connected to the stilling vanes is stopped. Scraper vanes on the

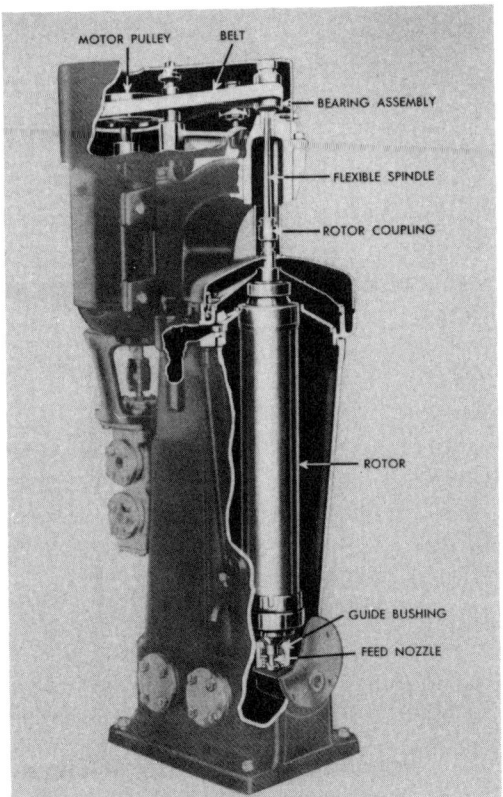

*Figure 4.14 Tubular centrifuge.
(Photo courtesy of Pennwalt Corporation).*

outside of the stilling vanes break through the settled solids and the turbulence created by this discharge scrubs clean the remaining solids in the machine. Approximately twice the volume of the machine is required to dump the solids out of the machine. The controller then restarts the machine through another cycle. All the valves are operated by the controller.

This centrifuge was developed to replace conventional clarifiers. It is very effective on streams containing less than 1,000 ppm. It is limited to low solids concentrations because of the limited solids-retention capacity. It is built in two sizes, a series 2500 and a 5000 series. It handles capacities of 10 m^3/hr (45 gpm) up to 250 m^3/hr (1,000 gpm). The centrifuge normally operates at 2,000–3,000 g's.

THEORY

Filtration Centrifuges

The theory of filtration centrifuges is not highly developed and attempts to

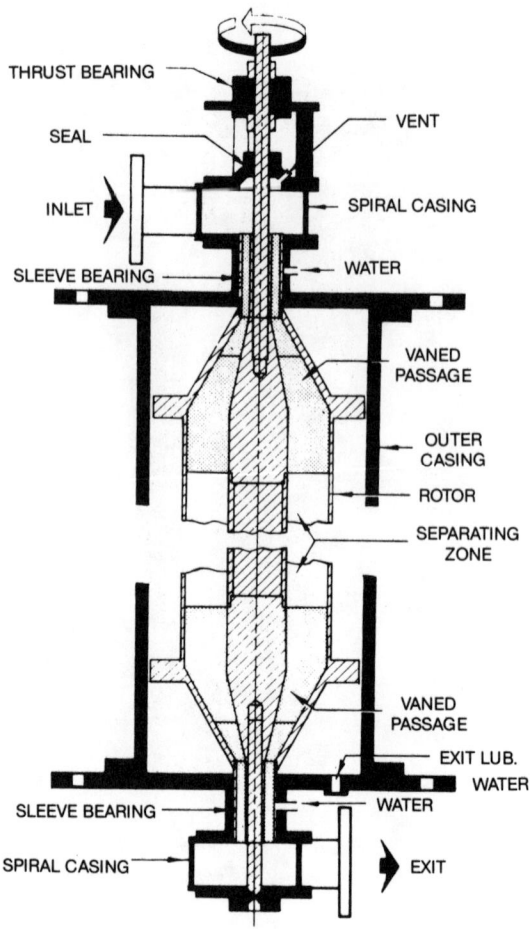

*Figure 4.15 Vortex Clarifier
(Illustration courtesy of Bird Machine Company, Inc.).*

apply it have met with limited success. Filtration under centrifugal force is more complicated than filtration under gravity because of the differences of the force levels on both the liquid being driven through the filter bed and the compression of the filter bed due to the increased gravitational forces through the filter cake. Cake filtration theory is normally described by the following equation, which does not allow for the compressibility of the filter cake:

$$V = \sqrt{\frac{Pt}{\mu^{\infty}}}$$

where V = volume of filtrate per unit area
       P = differential pressure across the filter cake
       t = time allowed for filtration
       μ = viscosity of liquid
       ∞ = specific cake resistance

This equation is used in the sizing and scaleup of vacuum filters. It is helpful perhaps in centrifugal filtration for defining directions only, and not in making quantitative calculations. The capacity of the centrifuge will vary proportionately as the square root of the solids concentration of the feed and inversely with the square root of the viscosity of the liquid and specific cake resistance. The specific resistance of the cake is governed by the particle size and shape. The moisture of the discharged product will be influenced mainly by the surface area of the material being dewatered; the greater the surface area, the higher the moisture. As stated previously, filtration centrifuges are evaluated on the basis of empirical laboratory tests, and scaled up on the basis of scaleup factors developed by the equipment manufacturers. These tests will be further discussed under the section entitled Laboratory Evaluations.

Sedimentation Centrifuges

The theory for Sedimentation Centrifuges has been more highly developed, but again has great limitations. The theory has been developed from gravity sedimentation theory and Stokes law. This has been developed for settling tanks and then further developed for use in Sedimentation Centrifuges. In settling tanks, Stokes law has limitations because it cannot allow for hindered settling that occurs as the solids concentrrate. In centrifuges the volumes are considerably more limited, and thus hindered settling takes place almost immediately; this greatly affects the value of theoretical calculations. Their usefulness is in determining scaleups for like types of centrifuges, which are operated under similar conditions.

One development from Stokes law on the theoretical settling area of centrifuges as applied to continuous horizontal helix types is the Sigma formula (1).

Without going into the development of the formula, the formula is as follows:

$$\Sigma = \frac{\pi b \omega^2}{2g} (3r_2^2 + r_1^2)$$

$$\Sigma - M^2 \text{ (ft}^2\text{)}$$

(of gravity settling tank or equivalent sedimentation
characteristics to centrifuges)

where r_2 = radius of inner bowl wall
       r_1 = radius of retained liquid surface
       ω = rate of rotation radians/sec
       b = length of cylindrical bowl
       g = gravitational constant

The Sigma calculations are based on the configuration of the bowl as contained in Figure 4.16. It will be noted that we only have used the active clarifying length or the length of the bowl from the feed point to the point of discharge. This calculation is quite similar for that of a Tubular Centrifuge. Other developments of the Sigma formula have used the conical section of the bowl also, but this has not been included as it is not part of the active clarifying volume of the centrifuge.

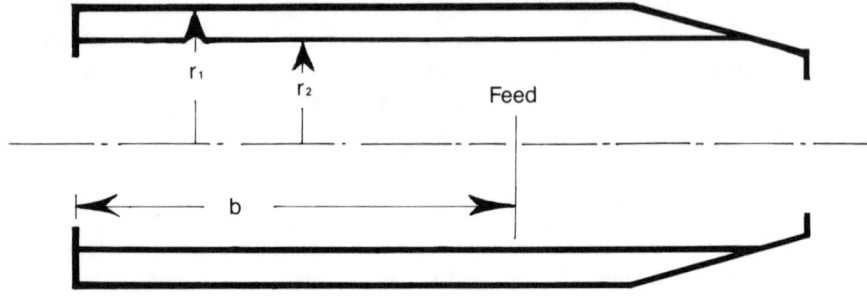

r_1 = radius to pool depth
r_2 = radius of inside of bowl
b = distance from point of feed to end of bowl

Figure 4.16 Sigma calculations solid bowl centrifuge.

The Sigma formula for a disc centrifuge may be expressed as follows:

$$\Sigma = \frac{2\pi (N-1)(r_b^3 - r_a^3)\omega^2}{3 \, g \, \text{ton} \, \theta}$$

where N = number of discs in the stack
       r_b = inner radius of disc stack
       r_a = outer radius of disc stack
       θ = half included angle of disc

The value of Sigma for commercial centrifuges will vary from 100 m² (1,000 ft²) for commercial basket centrifuges to over 10,000 m² (100,000 ft²) for large-disc centrifuges. The efficiency of the Sigma calculation will vary from one type of centrifuge to the other and is not a measure of comparative performance of the various type centrifuges. This is more of a function of the proper application of the equipment to the liquid-solid separation problem. To reinforce the statement, Sigma should be compared only to similar type method of operation, etc. The author showed capacities of solid bowl centrifuges with similar Sigma's had their capacities varied as much as 2:1 (2).

LABORATORY TESTS

The engineer faced with a separation problem can perform some simple laboratory tests to determine what levels of performance are possible. Simple equipment for these tests may be available in many laboratories. In addition, most equipment manufacturers have test laboratories that have extensive testing facilities available at modest prices or no charge, where they will evaluate small samples and recommend the most suitable equipment and expected performance.

All pertinent information regarding a separation problem should be assembled before conducting any evaluations. A typical "solids-liquid separation application data sheet" is shown in Figure 4.17. This data sheet will be useful in talking with equipment manufacturers, as it contains most of the information they would be looking for to provide preliminary recommendations on equipment.

After assembling these data, it is suggested that the engineer refer to Table 4.1 and determine what type of equipment would seem to best suit the separation problem. Then some simple laboratory tests can be conducted to determine preliminary performance levels.

The first test described will be for filtration-type centrifuges. For these tests a 4-in. diameter "pulp"-type centrifuge is required with a perforate basket. These are made by many of the laboratory centrifuge manufacturers. The basket should be lined with about a 250-μ (60 mesh) wire. Samples of the feed slurry should then be prepared, and enough slurry put in the centrifuge to form about 12—16-mm-thick (½—¾ in.) cake. Then a series of tests should be conducted at speeds to provide a centrifugal force of 500 and 1,000 gravities, and dewatering the cake at say 15 seconds, 30 seconds, 60 seconds and 2 minutes. This will produce a family of curves similar to that contained in Figure 4.18. Also analyze the percent solids in the effluent to determine approximate levels of effluent clarity.

From the above curve, the range of values that can be achieved with centrifuges can be seen. The retention times under 60 seconds are often achievable on continuous centrifuges. Longer retention times probably will only be achievable in batch centrifuges.

From this point there is considerable information to discuss the performance with equipment manufacturers. Tests on pilot plant equipment can be arranged in the equipment manufacturers' laboratory. Most equipment manufacturers have a pool of rental pilot plant equipment that can be employed in the engineer's production or pilot plant if this is more desirable.

Product Data

Company _____
Address _____ City _____ State _____
Plant Address _____ City _____ State _____

Process Data

Feed Solids Composition: _____
% Susp. solids (ss) _____ _____
Sp. gr. _____ _____
Viscosity _____ (cp) at _____ (°F) ☐ Slimy   ☐ Soft
Process ☐ Crystalline ☐ Fibrous
Tmp. _____ (°F) Pressure ___ (PSIG) ☐ Abrasive ☐ Amorphous
Rate _____ (GPM) _____ Sieve _____ % On _____ Cumul. %
_____ (lbs/hr ss) + _____ M _____
Valuable Component(s) _____ + _____ M _____
BOD or COD (for waste) _____ + _____ M _____
Type (Origin) of waste _____ + _____ M _____
Mother Liquor Sp. gr. _____ Shape _____
Composition _____ Ash Content _____
Dissolved: Solids _____ % Bulk Density _____
      Impurities _____ %
Sp.gr. _____ at _____ (°F) pH _____

Processing Requirements
Max. liquid content in solids _____
Max. impurities in solids _____
Max. ss in clarified ML _____
Material of const. _____
Disposition of cake product _____
Disposition of Mother Liquor (ML) _____

Flowsheet

Comments _____
_____
_____

Figure 4.17. Solids/Liquids separation application data sheet.

"Pulp" Centrifuge Dewatering Tests

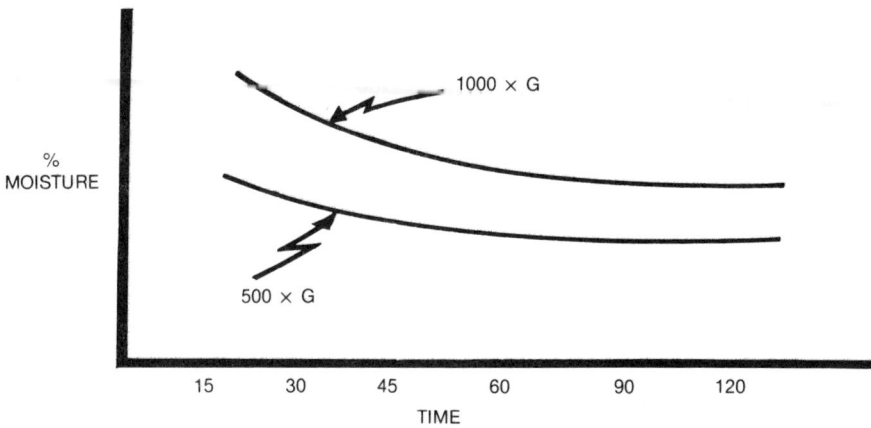

Figure 4.18 Perforate basket centrifuge tests.

It might be mentioned here the importance of properly selecting samples for laboratory evaluations. The procedures of predicting performance of centrifuges in a laboratory is quite accurate. However, it is important to select samples that are truly representative of the materials to be encountered in the production plant. ANalysis should be repeated on samples on which tests are conducted to make sure that the particle size or shape has not changed during shipment. This could greatly affect the anticipated performance of equipment. When selecting samples of coarse solids from tanks, the solids tend to stratify in the tank. Samples can be drawn that may be considerably coarser or finer than what would be represented by a homogeneous mixture. The important task of proper sample selection is often relegated to the least knowledgeable person in the plant. This often provides a very poor basis for a very significant capital investment.

For sedimentation-type centrifuges, tests can be conducted in a laboratory test tube-type centrifuge. The residence time in most continuous centrifuges is in the range of 5–120 seconds. A good procedure for conducting laboratory tests is to spin a sample in a laboratory centrifuge for periods of 15, 30, 60 and 90 seconds. On a data sheet, note the amount of settled solids along with the time centrifuged. Additionally, it is helpful to determine the degree of firmness by probing the bottom of the tube. Observe whether the settled material is packed firmly or is relatively fluid. If the material is packed firmly, the supernatant may be decantered off and the sedimented material analyzed to obtain the liquid content of the sludge. This will approximately provide a very rough estimate of the product discharged from a sedimentation centrifuge.

If the material settles rapidly and relatively firmly, it probably is a good candi-

Chapter 4. W. F. White

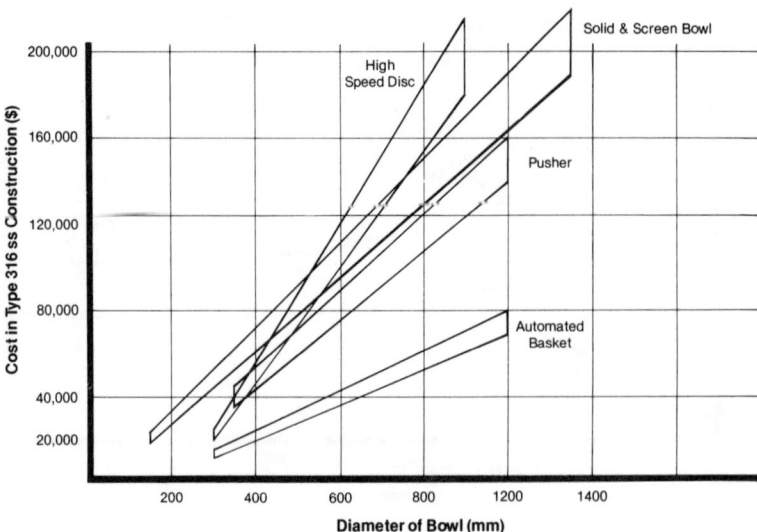

Figure 4.19 Cost of centrifuges used in the chemical process industries.

date for a continuous solid bowl centrifuge. If the settled solids are completely fluid, it may be a better candidate for a disc-type centrifuge. Again these type judgments are better made by a professional experienced in the field. This experience can be obtained by consulting with the equipment manufacturers of this type equipment. They will conduct these tests in their laboratory. It is often worthwhile to visit these laboratories and observe the tests being conducted.

Sedimentation centrifuges are often assisted by the use of chemical flocculation. This is particularly true in waste centrifuge applications. Flocculation studies are an art in themself. Assistance on this aspect can be obtained from the major manufacturers of high-molecular-weight polyelectrolytes such as Allied Colloids Inc., American Cyanamid, Dow Chemical, Hercules Inc., Nalco Chemical, Rohm & Haas Company, etc.

As stated previously for filtration-type centrifuges, the next step is to arrange for tests on pilot plant-size centrifuges.

In conclusion, Figure 4.19 may be referred to for a preliminary estimate of costs for the centrifuges only (no installation, electrical or accessory equipment) in Type 316 SS construction based on 1978 costs. Costs will vary depending on requirement for sealing, pressure and accessory equipment desired. Equipment manufacturers should be consulted for more definitive estimates.

REFERENCES

1. Perry, R. H., *Chemical Engineers' Handbook*, 4th ed. (New York: Mc Graw-Hill Book Co., 1963), p. 92.
2. White, W. F. "A Centrifugal For Industrial Wastes," *Chem. Eng. Prog.* (June 1969).

CHAPTER 5

LIQUID PRESSURE FILTERS

ALFRED TRUMPLER
Trumpler-Clancy, Inc.
Hamburg, NY

INTRODUCTION

Terminology & Definitions

1. Admix or Body Feed: The continuous addition of a filteraid to a liquid being filtered to improve filtration rate or clarity.
2. Blinding: Stopping of fluid flow through any filter media or septa because the pores are becoming blocked, plugged or clogged.
3. Colloidal: Extremely finely divided and dispersed solids in suspension in a liquid.
4. Efficiency: A trade practice term that involves the percentage of particles at any given micron level removed by a given filter medium compared to the total present.
5. Filter Cycle: A single continuous period of filtration; the period of filter operation between cleanings or replacements of media or septa.
6. Filter Medium: The porous substance that effects the retention of particles during filtration.
7. Filter Septum: The support device for a filter cake — wire screen, cloth, paper, etc. — specifically not the medium performing the filtration.
8. Filtrate: The liquid that has passed through a filter; filter effluent.
9. Micrometer (μm) (formerly micron): 1/25,000 of an inch (0.001 mm.).
10. Permeability: Capacity of a filter medium to permit fluid flow; the reciprocal of resistance.
11. Porosity: A measure of the void content of a filter medium or cake; expressed usually as a ratio of open space to total volume.
12. Pressure Differential: The hydraulic pressure difference between the inlet and outlet of a filter, indicating the actual pressure drop across the filter medium or septum, plus accumulated solids, plus any other internal hydraulic losses.
13. Retention: Capacity of a filter medium to retain suspended matter.
14. Slurry: Mixture of suspended solids and a liquid.
15. Turbidity: Marred clarity condition in liquid caused by presence of colloidal or larger solids or semisolids.

Chapter 5. Alfred Trumpler

SELECTION PARAMETERS

The filtration of industrial liquids is a vastly comprehensive and complicated area of concern for the process designer and engineer because there are so many possible filtration methods and equipment types from which to choose. One must only look under Filters, Liquid in major buyers' guides such as the *U. S. Thomas Register* to become aware of the hundreds of American manufacturers of equipment, media and materials and the unbelievable variety of possible approaches to almost any given filtration problem (Figure 5.1).

Figure 5.1 A small portion of the line of hardware, cartridges and media available from just one U.S. filter manufacturer. (Courtesy AMF Cuno Div.).

Although water and industrial liquids are filtered in gravity and vacuum filters as well as pressure filters, gravity filters are limited largely to high-capacity municipal sand and gravel filter beds, and a few specific industrial oil and coolant applications, and will not be covered in this chapter. However, vacuum filtration is a very major part of today's industrial filtration technology and practice and is covered in a separate chapter.

Pressure filters are by far the most widely used by industry today for removal of various amounts and types of suspended solid and semisolid impurities and contaminants from a great variety of liquids. In past years the main purpose of a filtration step was to remove an undesirable contaminant from a desirable or valuable liquid, and the contaminant had to be disposed of as a waste material in some form. Today, with heavy emphasis on liquid pollution, many filtrations are performed on wastewater and process liquids so that the liquid and the solids may both be disposed of in an acceptable, nonpolluting manner. Also, there are many industrial filtrations, in which the solids are the desirable or valuable product and the filtrate is discarded to waste. In some cases, both filtrate and solids are used in some way and neither is disposed of as a waste product.

It should be apparent to the process designer and engineer from this brief introduction that proper application is the crux of the problem in selecting pressure filters. It would require an entire textbook to go into sufficient detail to fully evaluate and discuss all the fine points of every available type of pressure filter, filter media and septa on the market today, so this discussion will be limited to providing general comparative information and application guidelines for selection among the most commonly used types (Figure 5.2).

Among the most fundamental selection parameters that must be considered are:

1. reslurried versus dry cake (solids) disposal;
2. high solids versus low solids;
3. filteraids versus no filteraids;
4. surface versus depth filtration;
5. nominal versus absolute particle removal requirements;
6. throwaway versus reusable media; and
7. capital cost versus operating cost.

Reslurried Versus Dry Cake (Solids) Disposal

In some instances, the methods of filtration and the types of equipment that can be considered for use are seriously limited by this parameter, particularly with waste disposal problems being so important today. There are filters in which the contaminants are discarded as part of disposable filter media. Some filter types described in this chapter can only be cleaned by washing the contaminants from the filter septa by sluicing or backwashing (*i.e.*, reslurrying the solids in water or the liquid being filtered) and, therefore, are not suitable for use where restrictions prohibit sewer discharge and where there is no waste treatment system in the plant to handle and treat such wastes.

Other types are designed to provide only a "dry" cake (this term generally refers to a filter cake in the range of 35–70% solids by weight), which is normally

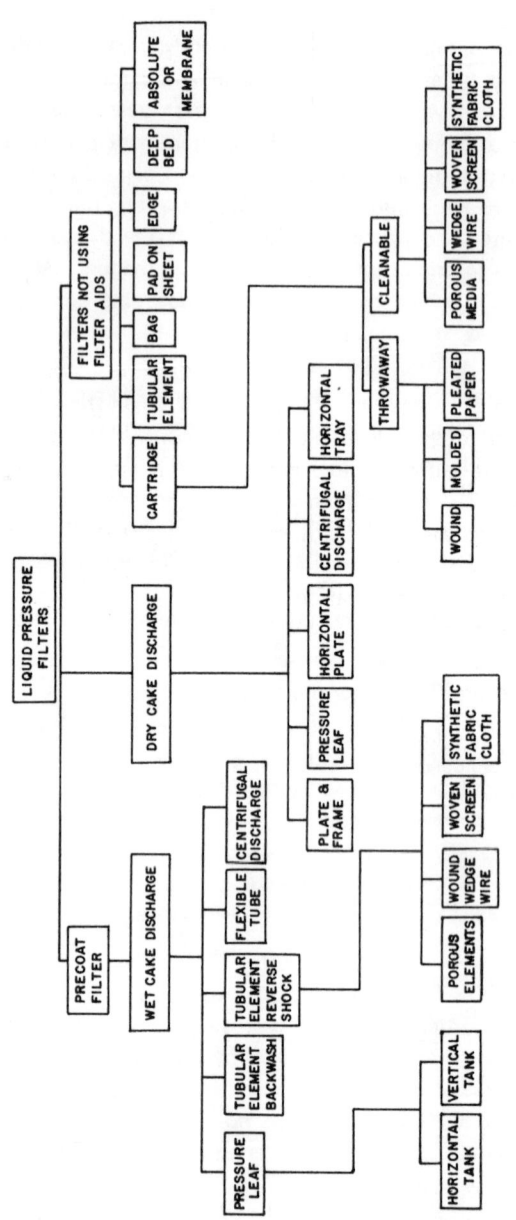

Figure 5.2 Schematic chart of liquid pressure filters covered in this chapter.

disposed of as a waste product in trash containers or as landfill. Still others can provide either dry cake discharge or a reslurried cake, usually by manual or automatic sluicing (Figure 5.3).

Figure 5.3 Dry filter cake dropping from pressure-leaf filter into hopper during vibrator operation. (Courtesy Industrial Filter & Pump Manufacturing Company.)

High Solids Versus Low Solids

Typical industrial filtration applications range from a few parts per million of unwanted suspended solids to slurries with 15–20% solids by weight. Obviously, a completely different type of pressure filter is required to handle each of these extremes. In the most common range of 50 ppm to 3% by weight, however, there are many gray areas of possible overlapping of different filter types, and many considerations are usually required for proper filter selection.

Filteraids Versus No Filteraids

Commercial filteraids will be discussed in some detail in the section "Filteraids and Their Application," p. 113, but let it sufice to say here that filteraids are materials that are relatively inert, insoluble solids used in a powder form to provide extremely fine filtration, higher filtration rates and longer filtration cycles than would be possible without them. The use of filteraids is generally required in applications where (1) solids loading is above 50 ppm, (2) where solids are of a compressible, slimy or gelatinous nature, and (3) where extremely fine particles in the submicronic and colloidal range must be removed.

The entire range of pressure filters is divided into types that normally use filter-aids (often called precoat filters) and those that normally do not. The main differences between precoat filters and other types relate to the use of filter septa versus filter media in their construction, space for filter cake accumulation, and methods of cleaning and contaminant disposal.

Surface Versus Depth Filtration

In true surface (sometimes called screen) filtration, all particles removed from a liquid are retained on the surface of a porous medium by a sieving mechanism and do not penetrate into the medium at all. In true depth filtration, a filter medium of random porous structure and of some thickness is made by various means to entrap particulate matter *within* the medium as well as on its surface.

Each method has certain advantages if properly applied. In removal of slimy, gelatinous solids, like yeast or algae, a simple porous filter surface will blind quickly and require replacement or cleaning after a relatively low volume of solids has been deposited. For this type of application, depth filtration would normally be far superior in terms of throughput and filter cycle length.

Conversely, in removal of crystalline particles like sand or salt crystals, a highly porous surface medium would be quite adequate for particle removal and would exhibit minimal blinding tendency, providing relatively high throughput and long filter cycles.

In actual practice, most filter media are neither true surface media nor true depth media, but a combination of the two, retaining some particles within the media and others on the surface. Or, when a final absolute (see below) surface medium is needed, a depth-type prefilter is often used to remove the bulk of the solids. In any case, the designer and engineer must be aware of this important parameter in selecting equipment and must in many cases perform tests to determine which type of media will provide lowest operating cost as well as desired removal efficiency, and then select a type or types of filter that offer the media desired.

Nominal Versus Absolute Filtration

This parameter is perhaps one of the most important from the standpoint of initial and operating cost of pressure filters. Nominal micronic particle removal capability assigned by various manufacturers of filter media means just that: particle removal rating is approximate and varies with conditions. What then does a nominal rating mean? If a manufacturer of nominally rated filter cartridges, for example, is pinned down, he may say only that with a specific laboratory test dust (standard contaminant) in a certain liquid, with a specific flowrate and pressure drop, on a once-through test, his cartridge is 90% or 92% or 98% efficient in removing all particles larger than X micrometers and, therefore, has a nominal rating of X micrometers.

However, with another liquid, containing different contaminants, and at different flows and pressures, there is no assurance whatever to the user that the same efficiency at the X micrometer level will apply, nor is there any certainty as to the maximum size particle that can be forced through that cartridge under various operating conditions.

It is apparent, therefore, that nominal micrometer ratings applied by most manufacturers to their various filter media are little more than comparative guides to particle removal effectiveness, and either laboratory testing or production trial-and-error methods are needed to determine what is both effective and economical for a given service.

One may wonder why the vast majority of filter media used by industry today are so inexactly identified in terms of performance capabilities. However, it is easily explained by noting that most filtration users have only the vaguest idea of what they want or need in a given process; and when the nature of the need is inexact, the products serving that need may also be quite inexact and still serve their purpose very well.

Absolute filtration is an entirely different matter. Theoretically, the term "absolute" should mean just that — and if a user demanded a 1 μm absolute filter medium, he should be guaranteeed that 100% of all particles larger than 1.00000 μm would be removed when challenged by that filter medium. In practice, most absolute ratings are qualified either by a "±" after the maximum pore size or by an efficiency rating such as 99.999%.

However, these slight qualifications of the absolute ratings of various filter membranes and other media do not detract from their effectiveness in controlling turbidity in critical liquids where sterility of yeast and/or bacteria is required.

The important considerations on the part of the engineer or designer are the economic factors involved in selecting types of filters that offer nominal filtration at relatively low initial and operating costs, as against absolute filters and media, which are quite costly, both in terms of initial hardware and replacement media. A decision to use pressure filters much more exacting in their capabilities than the particular application really *needs* can be considerably more costly than is necessary.

Throwaway Versus Reusable Media

The ideal filter medium is one that is inexpensive, is available in a wide particle removal range, including the submicrometer range, and never blinds or, at worst, provides long life before blinding and requiring cleaning or replacement. Unfortunately, there is no such thing, and selection must be made between throwaway and reusable media from a number of standpoints.

Throwaway media — such as many types of paper and fiber cartridges, bags and paper discs — are ideal in terms of ease of contaminant disposal, filter downtime between cycles, and low labor and convenience. However, cost per gallon of liquid filtered must be a prime consideration, and, in many instances, the unit filtration cost with throwaway cartridges is prohibitive. Therefore, filters with relatively permanent, cleanable and reusable media, such as woven fabrics, wire cloth, wound wire and porous materials, must be used.

The big unknown with reusable media, however, is the cost of operating labor and maintenance to keep reusable media reusable. All filter media have blinding tendencies of various degrees. In one application, certain filter media may have considerably less blinding tendency than others, and the difficulty, without testing in actual service, is to predetermine which media will provide the longest trouble-free life, and require the least maintenance to return the media to something near original permeability at reasonable intervals.

Capital Cost Versus Operating Costs

There is seldom only one answer to a filtration problem and quite often a decision must be made between two equally effective methods of handling a specific application.

One method may well involve a type of filter the initial cost of which is fairly high, but the day-to-day operating cost of which is relatively low (for example, vacuum or pressure precoat filters); the other method may require a type of filter with relatively low cost but rather high operating cost (for example, throwaway cartridge filters).

IMPORTANCE OF FILTER SEPTA

Once it has been established by a consideration of the above parameters that the application in question is a relatively high solids filtration (above 50 ppm), that the contaminants are of a nature prohibiting surface filtration, and that filteraids will probably be needed, then consideration of the filter septa must be given.

Actually, since the septa perform no filtration but are merely the support for a filter cake, whatever it may be, the type of septa selected is dictated by cost, corrosion resistance, compatibility, cleaning and maintenance labor, blinding tendencies, cake release characteristics, strength, porosity and resistance to flow. Also, not all types of filter septa can be used in all types of precoat filters, so often the consideration of septa must be concurrent with consideration of the various types of possible filters (hardware). More will be said about various types and forms of filter septa under the pressure filter types discussed in the section "Filters Designed to Use Filteraids," p. 116.

IMPORTANCE OF HARDWARE

Were it not for the fact that selection of suitable filters (hardware) is tied to a large degree to selection of filter media and/or septa, the task of picking the best filter out of the hundreds of types available would be next to impossible.

However, once sizing and capacity requirements begin to point the direction, media or septa are selected and the other parameters considered, most filter types begin to fall by the wayside as obviously unsuitable, and the final selection becomes somewhat limited.

It is to the evaluation and comparison of features and characteristics of the most widely used pressure filters in today's market place that this chapter is devoted.

FILTERAIDS AND THEIR APPLICATION

No consideration of pressure filters could be complete without a discussion of filteraids and their importance in many industrial filtrations (Figure 5.4).

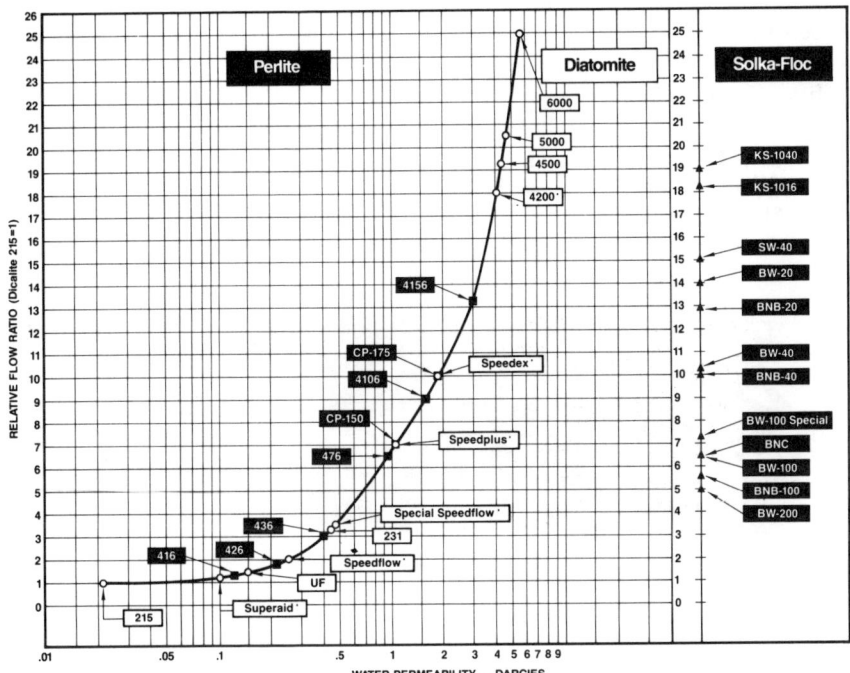

Figure 5.4 Relative flowrates of various perlite, diatomite and cellulose filteraids. Filtrate clarity is normally inversely proportional to flowrates. (Courtesy Dicalite Division, Grefco, Inc.)

A filteraid is a powder or fibrous material that is essentially inert, essentially noncompressible, and highly porous and permeable, and serves several purposes in filtering many liquids. Removal of certain types of contaminants cannot be achieved economically with the use of most types of filter media, since they either do not remove submicronic and colloidal particles, or they tend to blind too quickly for practical use.

Filteraids are usually used in two ways. First, since relatively coarse septa such as wire cloth, woven fabrics, porous media or filter paper are used only as a base or support for the filter cake, a precoat layer of the filteraid, about 1/16–1/8 in. (0.16–0.32 cm) thick, is initially formed on the septa by recirculating a slurry containing filteraid through the filter until all the precoat material has been deposited on the septa.

This precoating procedure serves the dual purpose of establishing the clarity level which the particular type and grade of filteraid provides, and also protecting the filter septa from direct contact with the contaminants which may be slimy, gelatinous and compressible, and would otherwise tend to blind the septa.

However, in many instances, this precoat layer of filteraid would soon be blinded by a thin layer of the contaminant after the process filtration had begun, were it not for the continuous addition, or admix of filteraid to the liquid ahead of the filter (Figure 5.5).

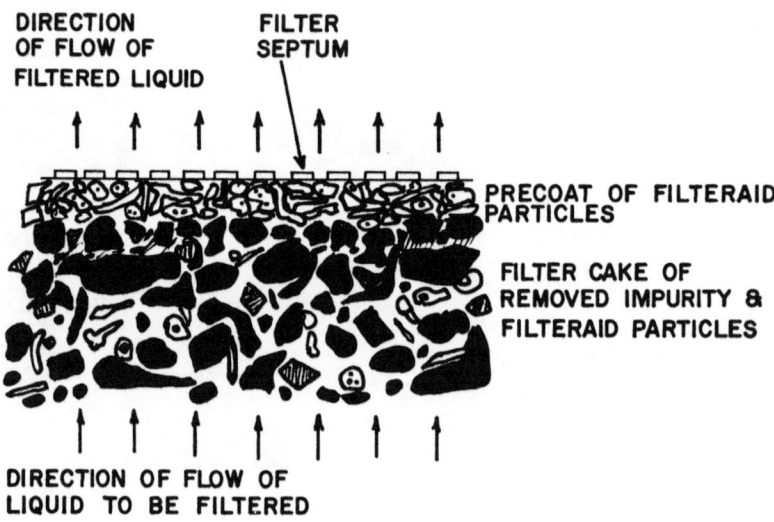

Figure 5.5 Diagram showing how filteraids are used both as precoat and admix in a typical process filtration. (Courtesy Kenite Division, Witco Chemical Company.)

Just a word on filtration theory at this point may be helpful in understanding the use of filteraids as admix. Basic filtration is concerned with liquid flow through a unit of filter area per unit of time. Expressed as a formula, this is:

Process Equipment Series Volume 1

$$\frac{dV}{d\Theta} = \frac{\Delta P}{R_c \times \mu \times L}$$

where V   = volume of liquid
       Θ = time
       ΔP = pressure drop
       R_c = specific cake resistance
       μ = viscosity
       L   = cake thickness

It is apparent from this simplified formula that since filtration rate is inversely proportional to cake resistance, anything that can be done economically to reduce cake resistance (increase cake permeability) will increase rates, and this is exactly why filteraids are used as admix, because they tend to increase cake permeability, often quite considerably.

Although the optimum grade and amount of admix can best be determined by laboratory testing, a rule of thumb of equal parts of filteraid to contaminant with crystalline solids, up to a 5:1 ratio with slimy, compressible solids, will normally maintain reasonable permeability of the increasing filter cake (Figure 5.6).

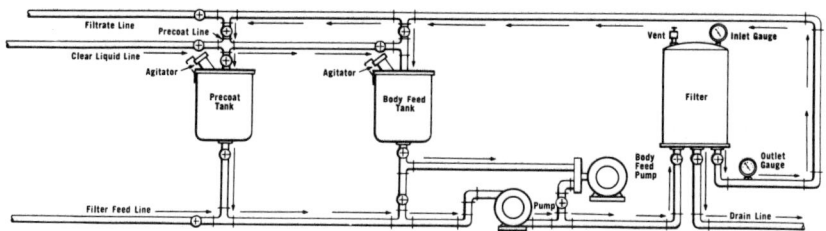

Figure 5.6 Schematic sketch of a typical filtration system used with precoat filters employing body feed filtration (Courtesy Celite Division, Johns-Manville Corporation.)

Three types of filteraids are most widely used today: diatomaceous earth (diatomite), perlite and cellulose. Asbestos fiber has, for many years, been used because of its positive charge and certain other characteristics, but at this writing, certain government regulations have resulted in almost complete removal of this material from filtration in the USA.

Diatomite, which has been used as a filteraid for more than 50 years, is an amorphous form of silica, is free of organic matter, is essentially inert and is virtually nonadsorptive (Fig. 5.7). Perlite, a slightly more recent filteraid, is made from volcanic glass and is an amorphous form of fused sodium potassium aluminum silicate, also inert, nonadsorptive and free of organic matter. Because of its extremely light density and certain special characteristics, perlite is used more widely as a precoat for rotary vacuum precoat filters than for pressure filtrations, although an increasing number of pressure filter applications are using perlite today.

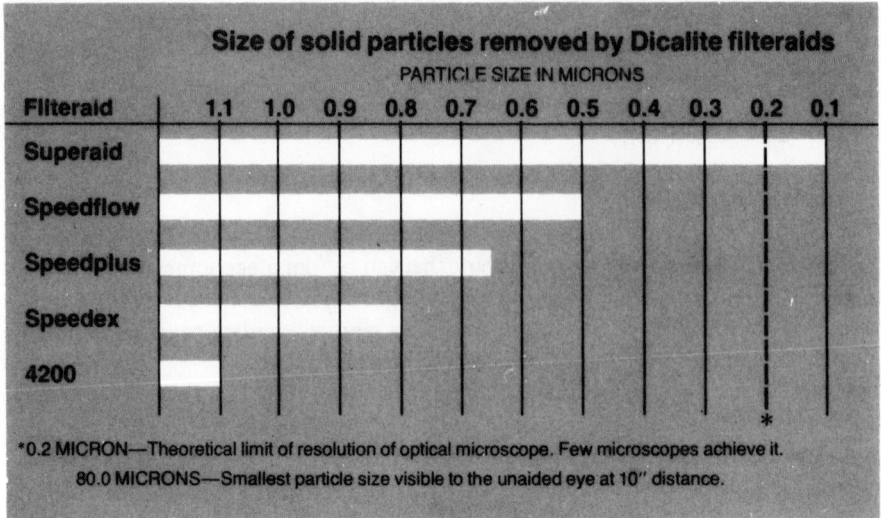

Figure 5.7 Chart showing the sizes of particles removed by five grades of diatomite filteraids. The data were established through the use of suspensions of uniform particles of known size. (Courtesy Dicalite Division, Grefco, Inc.)

Cellulose filteraids are refined, pure cellulose, essentially ashless and nonabrasive, and are generally used where fibers are essential or helpful (such as in precoating many wire cloth filter septa), and in applications where trace silica solubility or other impurities from the other filteraids cannot be tolerated. Cellulose filteraids are particularly helpful in providing more rapid precoating of certain septa, plus a degree of filter cake stability not otherwise obtainable. However, the much higher cost of cellulose filteraids (3 or 4 to 1) compared to diatomaceous earth and perlite must be considered in deciding on their use.

FILTERS DESIGNED TO USE FILTERAIDS (PRECOAT FILTERS)

All precoat filters are designed to provide space for a filter cake of varying thickness. This filter cake consists of the contaminants being removed, plus the filteraid or filteraids (often two types are used in the same filtration) used as precoat and as admix throughout the filtration cycle. When, at the end of the cycle, this accumulated cake is removed from the filter by reslurrying it, either by sluicing it from the septa with water or some other liquid, or by backwashing it with water or filtrate, this is referred to as wet cake or reslurried cake discharge.

When, on the other hand, the accumulated filter cake is mechanically compressed and/or dried in place by compressed air at the end of the filter cycle to a dryness of 35–70% solids by weight and is then removed from the filter and disposed of as a solid waste, this is referred to as dry cake discharge.

WET CAKE DISCHARGE FILTERS

Wet (reslurried) cake discharge filters commonly fall into two types: leaf filters and tubular element filters.

Pressure-Leaf Filters

Pressure-leaf filters are commonly manufactured in horizontal and vertical tank designs. In either case, they consist of a pressure vessel, generally designed to operate at up to the range of 60–100 psig (4.22–7.031 kg/cm^2) and contain a number of two-sided filter leaves mounted in a vertical position within the vessel.

Filter leaves are made in a wide variety of shapes, sizes and constructions. They may be round, square, rectangular or trapezoidal. They may be bottom discharge, side discharge or center discharge, but the one feature they all have in common is that they are made in a sandwich-type construction with two layers of some type of filter septum on the outside and a drainage member between the two septum layers, to provide channels for the filtered liquid to pass from the septum to an outlet channel. This drainage member, which also serves to some degree as a structural support, may be one or more layers of very coarse woven wire mesh (often 4 × 4 mesh), or expanded tubular slit screen metal. The entire sandwich assembly is normally bound at the outside periphery by a tubular channel, which, in the case of bottom drainage leaves, also serves as a collection passage to carry filtered liquid to the outlet nozzles seated in a discharge manifold (Figure 5.8).

In the case of center outlet leaves, the outside tubular channel is purely a means of sealing the edges of the leaves, and flow of filtered liquid is toward the center hub, which provides sealed passages into a central discharge manifold. In either case, the peripheral channel normally secures and seals the outsides of the leaves by a continuous weld, spot welds or by rivets spaced 2 or 3 in. (5.08 or 7.62 cm) apart all around each leaf.

The filter septa most commonly used for leaves are closely woven calendered wire cloth (usually stainless steel) and cloth bag coverings of many different filter fabrics and weaves. When wire cloth is used, it is part of the sandwich construction and is a permanent, integral part of the leaves (Figure 5.8). When fabrics are used as the filter septa, the basic leaf structure normally consists of just the one or more layers of drainage member, the outer channel and the outlet nozzle. The fabric bags are then fitted over the entire leaf, exposing only the outlet nozzles around which the bags must be tightly sealed.

Regardless of which septa are used, the main requirements of the septa are that: the opening size be small enough to permit buildup of a precoat of normal filteraids

Chapter 5. Alfred Trumpler

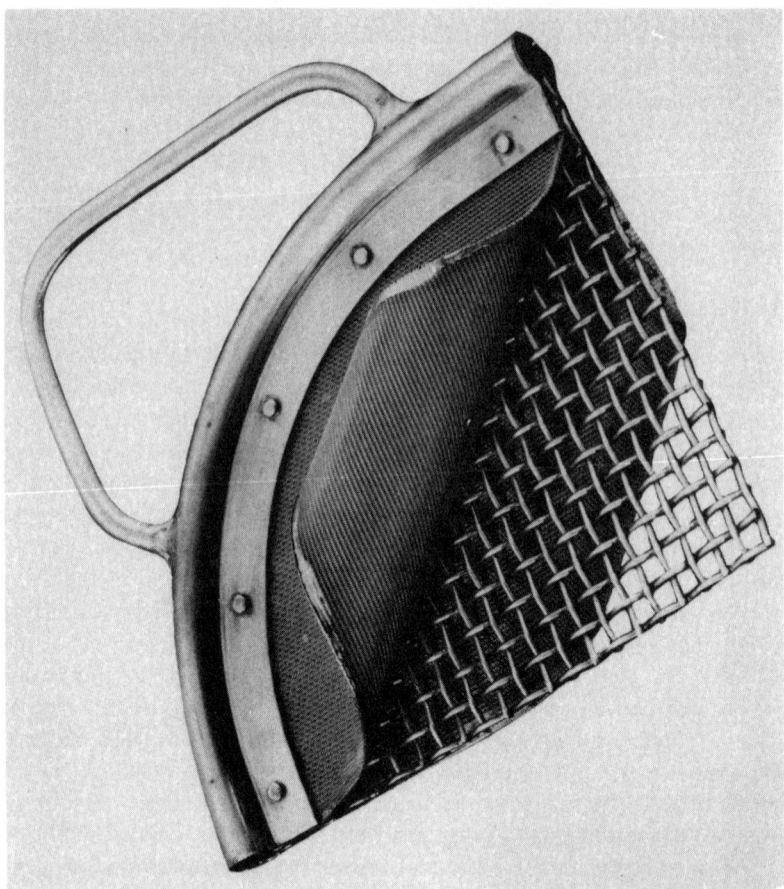

Figure 5.8 A cutaway corner of a typical riveted filter leaf utilizing two layers of 24 x 110 mesh calendared Dutch-weave wire cloth separated by a 4-mesh drainage member. (Courtesy Niagara Filters Division Ametek, Inc.)

within a reasonable (10—15 minute) recirculating period; (2) they have a minimum tendency to blind; and (3) they have good cake-release characteristics. Wire cloth most widely used is a 24 $\times$ 110 mesh calendered wire cloth made of stainless steel, monel, nickel or titanium (60 $\times$ 60 mesh and 80 $\times$ 80 mesh are not uncommon); fabrics are available in a great variety of materials (polypropylene, cotton, nylon Orlon®, Dacron®, Teflon®* and others), thicknesses and weaves. Selection is generally based on cost, fabric compatibility, permeability, strength, wear resistance and cake-release characteristics.

---

*Registered trademark of E. I. du Pont de Nemours and Company, Inc., Wilmington, Delaware.

Vertical tank pressure-leaf filters (Figure 5.9) are generally designed for wet cake discharge, either by manual or mechanical sluicing, but some manufacturers offer dry cake discharge versions, which will be discussed later. In this design, leaves are vertically mounted with bottom outlet nozzles fitted into a discharge manifold (Figure 5.9). Leaves are always stationary. After completion of a filter cycle, the filter is drained or blown out by air pressure, and the filter cake is either manually or mechanically sluiced (washed) from the leaves to sewer or a waste treatment system.

Figure 5.9 Interior of a vertical tank pressure leaf filter showing half of the filter leaves in position. (Courtesy Industrial Filter & Pump Manufacturing Company.)

With mechanical sluicing, either stationary, oscillating or rotating sluice headers are built into the filter cover and without opening the cover, the high-pressure sprays remove the filter cake and discharge it through the drain valve.

In this way, downtime for cleaning is shortened, but there may be some disadvantage in that a stubborn filter cake may not be completely removed and the next filter cycle may be entered without all the septa being cleaned. Also, one or more clogged or misdirected nozzles in the sluicing header may have the same results, and this is not known until such time as the filter is eventually opened for inspection.

Horizontal tank leaf filters have a number of design variations because each manufacturer has certain features he claims for his own design. Basically, these filters consist of cylindrical horizontal pressure vessels in which a sort of carriage assembly is positioned. This carriage assembly usually consists of the filter cover, a number of round or rectangular filter leaves, the outlet manifold in which the bottom outlet leaves are mounted, and whatever structure is needed for leaf positioning and to provide rigidity to the entire carriage. The entire carriage assembly rolls in and out on wheels, both at the inner end of the carriage and attached above the cover (Figure 5.10).

Figure 5.10 Horizontal tank filter, partially open, without filter leaves installed. Note internal carriage structure and sluicing header in top of tank. (Courtesy Industrial Filter & Pump Manufacturing Company.)

In the tank itself are tracks on which the carriage wheels are supported, baffles to properly distribute fluid flow and some type of sluicing arrangement. This mechanism usually consists of a sluicing header or headers positioned above the leaves, either stationary with wide angle nozzles designed to provide sharp cutting sprays to clean all filter leaf surfaces, or rotating or oscillating in some way to provide suitable cake removal. Other variations include top-mounted sluicing nozzles which move within the tank from one end to the other, cleaning off one leaf at a time, and dual oscillating sluice headers which oscillate 180° around the periphery of stationary bottom outlet round leaves.

One variation of the standard fixed tank design features a retractable tank (Figure 5.11) which is rolled away from a fixed head and leaf carriage assembly. However, since inlet and outlet connections are in the fixed head, internal hydraulics problems of flow distribution may exist in some designs.

Figure 5.11 Horizontal tank leaf filter, retractable tank design. (Courtesy Sparkler Manufacturing Company.)

Another variation of the horizontal tank filter is the use of rotating leaves (Figure 5.12) instead of stationary or fixed leaves. In this design the leaves are round, with center hubs mounted on a centrally located outlet manifold. The entire manifold and leaf assembly may be rotated slowly during precoating the filter and also during sluicing of the cake from the leaves by means of a fixed sluicing header.

If wet cake discharge is a suitable method of disposing of waste solids, a vertical or horizontal sluicing type pressure-leaf filter is surely one satisfactory way of handling filtrations where filteraids are required. Vertical tank units are available from roughly 20–1,000 ft^2 (1.86–92.9 m^2) of area and require relatively little floor space but considerable ceiling height for leaf removal; horizontal tank units are available from about 100–2,000 ft^2 (9.29–185.8 m^2) of filter area and can be utilized where floor space is available.

The horizontal tank units offer greater ease of inspection, since, in all except the

Chapter 5. Alfred Trumpler

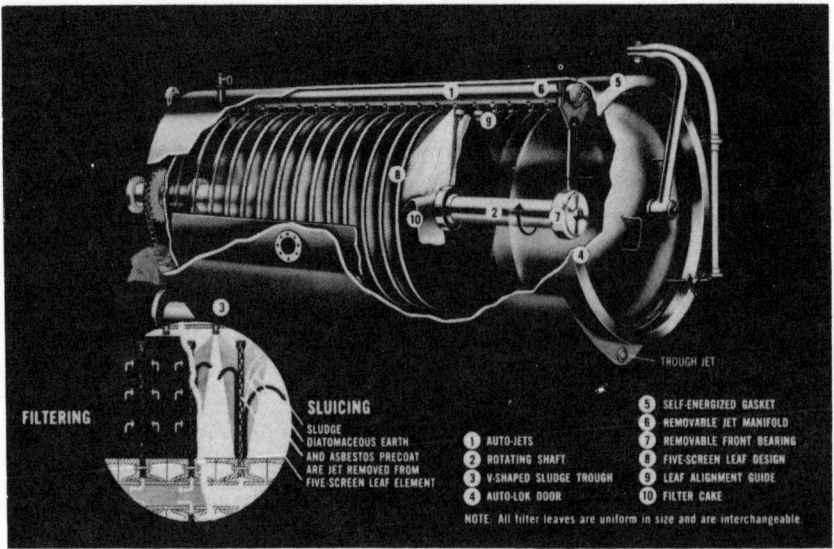

Figure 5.12 Rotating leaf pressure filter. Note sluicing header and center outlet manifold. (Courtesy U. S. Filter Corporation.)

rotating leaf designs, the entire leaf assembly can be exposed for visual inspection in a matter of seconds. In the vertical tank units, leaves must be removed individually from the top of the tank. Furthermore, sluicing the vertical tank leaves is more difficult, especially in larger sizes, because the distance from the sluice header to the bottoms of the leaves can be 5 ft (1.5 m) or more. With the horizontal tank construction, particularly with oscillating sluice headers and in the rotating leaf designs, no parts of the leaves are ever more than 2–3 ft (0.6–0.9 m) away from the sluice nozzles, so cake removal is likely to be more effective.

When filtering aqueous liquids, a sluicing leaf filter offers certain advantages in that cleaning of the filter between filter cycles with built-in sluice nozzles is easy and quick, requiring very little operator labor and reducing downtime to a minimum. It is generally unnecessary to even open the filter, except at certain intervals for general inspection of filter septa integrity and cleanliness and to check on the effectiveness of the sluicing arrangement.

There is, however, a greater likelihood of frequent internal blinding of the filter leaves because of the tendency of most sluicing nozzle spray patterns to drive a certain amount of the filter cake inward through the openings in the septa, particularly wire cloth, and gradually accumulate inside the leaves and completely block portions of the leaves until effective filter area becomes considerably reduced. With fixed bottom outlet leaves this tendency is minimal because this type of leaf tends to flush itself out internally and discharge solids through the leaf nozzles into the outlet manifold. (For this reason, it is always desirable to have a manifold drain open during the sluicing operation.)

However, the rotating leaf design and others with center or side outlets do not provide this self-flushing feature, and internal blinding may be a more serious problem; inspection and regularly scheduled (possibly weekly or biweekly) leaf cleaning procedures must be followed. Such procedures often involve circulation with, or soaking in, mild caustic or acidic cleaning solutions to free and, in some cases, actually dissolve these undesirable deposits.

Before leaving the subject of wet cake discharge leaf filters, it should be noted that these filters are *never* cleaned by backwashing (For our purposes, centrifugal discharge filters are not considered as leaf filters.) All common leaf designs are such that backwashing not only would be quite ineffective but also would provide a strong likelihood of leaf damage.

Tubular Element Backwash Filters

There are far more tubular element filters used for reslurried solids discharge than any other type, primarily because cylindrical filter septa are much more effectively cleaned by reverse flow of clean fluid (backwashing) than any other configuration.

Of the many types of filter septa used in tubular filters, the most extensively used are (1) porous ceramic, porous carbon and porous stainless steel, (2) wound wedge wire, (3) woven screen and (4) synthetic fabric cloth. Flexible braided wire and fabric mesh are discussed under "Flexible-Tube Filters" p. 130.

Porous Media — Porous tubes in various lengths are offered in molded ceramic materials in a wide range of permeability (Figure 5.13). These ceramic elements are relatively inert and not subject to corrosion problems, and they also are usable in temperatures up too 900–1000° F. (482–538°C). The higher-permeability materials have open areas up to 40% of total area. When used with filteraids, the elements are generally precoated as previously described, and filteraid admix may also be used if necessary. The accumulated filter cake is removed by backwashing at the end of the filter cycle.

The use of porous ceramic filters with filteraids offers the user extremely fine filtration levels. If properly precoated at the beginning of each filter cycle, the blinding tendency is minimal, but when blinding eventually does occur, the septa can usually be renewed by either acid or ultrasonic cleaning.

One disadvantage of porous ceramic filter septa is their brittleness. They tend to crack under high differential pressures or shock-type backwash procedures, improper assembly, and, of course, they will readily fracture if dropped or struck against a hard surface.

Tubular filter elements of porous carbon are also available from a few manufacturers. Their application and use as filter septa are similar to those of porous ceramic elements. They are generally quite brittle, crack easily and are used mainly

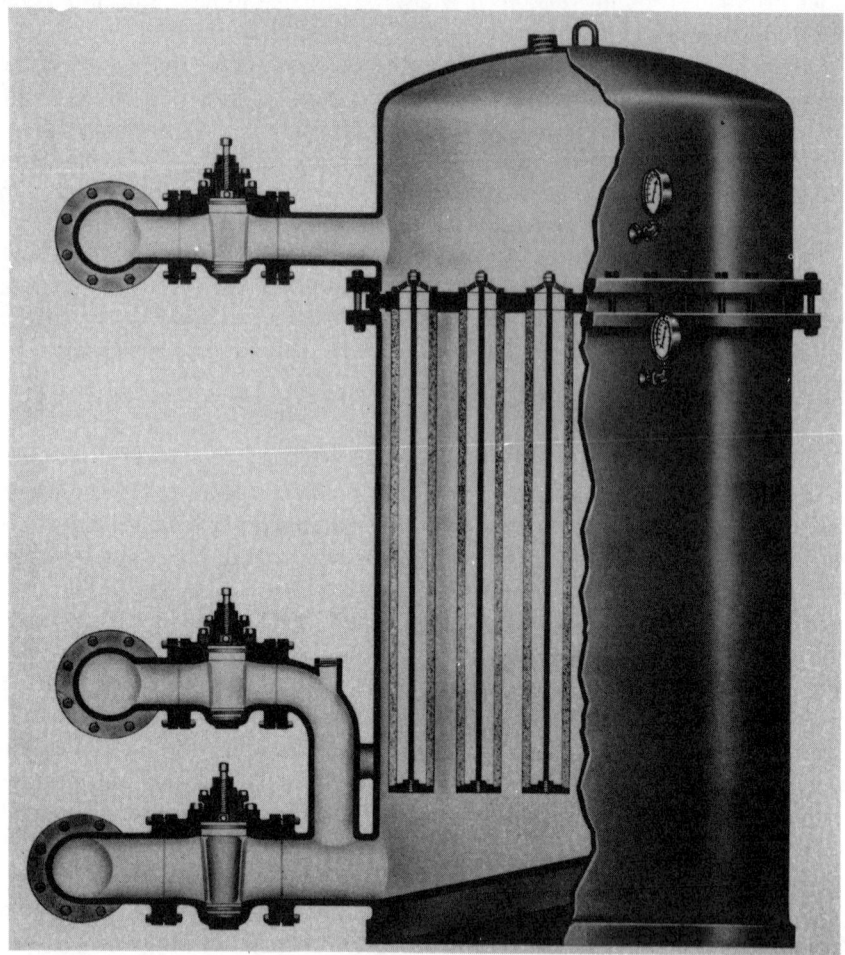

Figure 5.13 Typical porous ceramic filter cutaway. Compressed air in dome forces filtrate or water backwards through elements for cleaning. (Courtesy R. P. Adams Filter Co.)

in situations where product contamination, high temperature or chemical inertness require their use.

Thin-walled, porous stainless steel tubular elements are yet another form of porous septa used in some industrial applications. These elements are made by controlled sintering of stainless steel powders. Although they can be used with filteraids as precoat filters, the cost of these elements is so much higher than ceramic or carbon elements that they are seldom used in this way, although they are considerably stronger than other porous elements. Comments on backwashing and other cleaning methods made under ceramic and carbon elements apply equally to porous stainless steel.

Wound Wedge Wire — Tubular elements made of helically wound "wedge wire" septa are sometimes used both for trap filtration and for precoat filtration (Figure 5.14). Wound with triangular or trapezoidal wire around longitudinal rods and resistance welded at all joints, these elements provide a number of advantages over other septa.

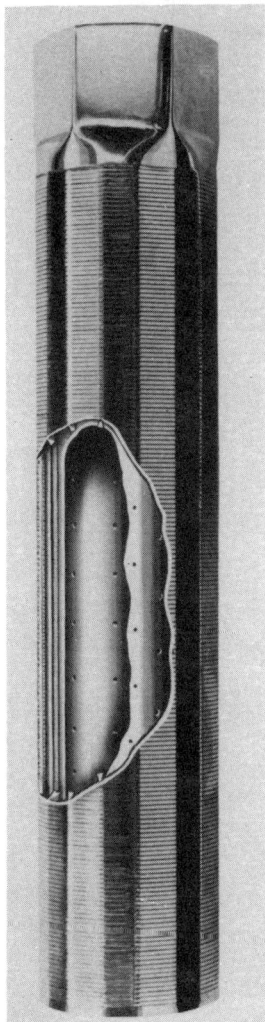

Figure 5.14 Wound wedge wire element cutaway. (Courtesy Croll Reynolds Manufacturing Co.)

Any filter septa meant to be cleaned by backwashing provide a serious concern to the process designer and engineer with reference to blinding tendency. Since the wire used in these elements is essentially triangular and smooth, with the apex pointing inward, the narrowest flow passage is at the outside surface, and the high backwash velocity at this point tends to clean the element thoroughly. Therefore, the blinding tendency is greatly reduced. One of the major manufacturers of filters utilizing this septum also offers means of adding compressed air to the backwash fluid to aid in achieving complete cleaning of the septum.

Furthermore, solids not trapped by the narrow slot openings at the outer surface of these elements tend to pass on through the element entirely because they encounter no corners or sharp or rough edges to stop them (Figure 5.15).

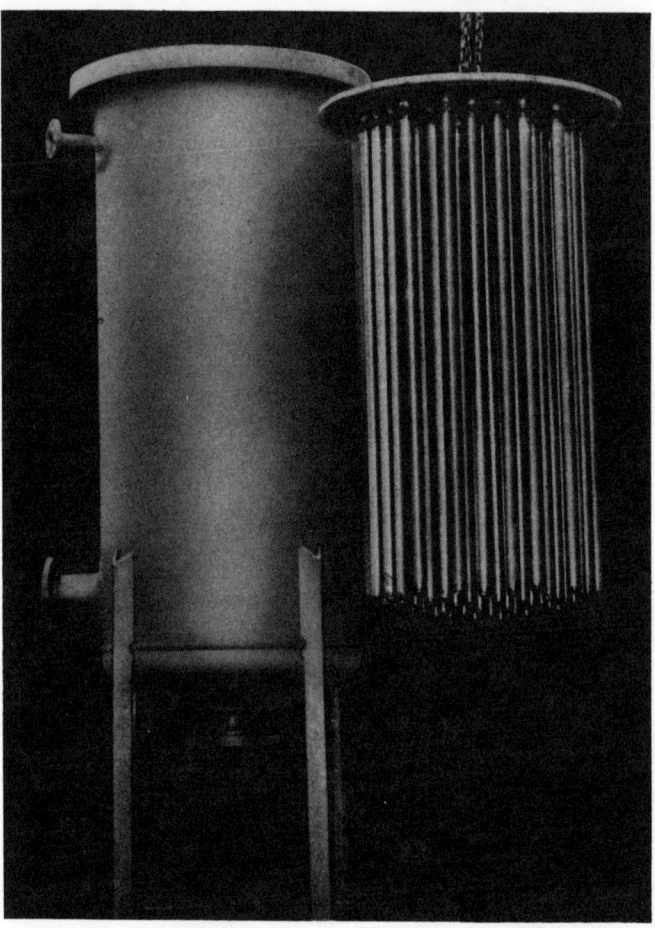

Figure 5.15 Vertical pressure filter with cartridge of wedge-wire elements suspended from tube sheet. (Courtesy Croll-Reynolds Manufacturing Co.)

Since the wire used in these elements is quite thick (generally more than 1 mm at the widest point of the section), these elements are extremely sturdy and rigid and are, therefore, difficult to damage. They can withstand high differential pressures and are also resistant to hydraulic and thermal shock, and warpage or distortion.

Wedge wire elements are available in carefully controlled slot openings from 0.001 in. (0.025 mm) (approximately 25 µm) to 0.0625 in. (1.59 mm) for fine straining. They are generally made of stainless steel but are available in any weldable material including certain plastics. Vertical precoat filters containing these elements are available with up to 1500 ft^2 (139 m^2) of filter area.

Disadvantages are relatively high cost and low open area in proportion to overall surface area.

Woven Screen (Cloth) — Woven screen or cloth filter septa are available from a large number of manufacturers and in a wide variety of materials and weaves. Woven metal wire cloth is available in stainless steel, brass, copper, monel, steel, aluminum, bronze, inconel, tantalum, Hastelloy and titanium. Plastic woven cloth media, made of monofilament nylon, polyester, polypropylene, Teflon and polyethylene are available in permeabilities from 7 µm to 5 mesh. Synthetic fabrics offer advantages of much lower cost and, in many cases, superior chemical resistance and reduced blinding tendencies; however, most are more sensitive to physical damage and elevated temperatures, and have limited solvent compatibility.

Weave types include plain or square weave, twilled weave, plain dutch weave, twilled dutch weave, and braided or basket weave. Each has its special characteristics with regard to strength, wire thickness, susceptibility to blinding, resistance to abrasion, open area, cake release characteristics, etc. It is not within the scope of this chapter to discuss in depth the technology of woven screens.

Fabric Cloth — Although some of the septa discussed under Woven Screens are made of synthetic fabrics, woven fabric *screen* and fabric *cloth* are quite different from each other. Historically, ordinary cotton duck fabric blankets were used as the filter septa on plate-and-frame filter presses for many years before woven wire cloth and woven synthetic filter fabrics for industrial use were even available.

Currently, woven fabric cloth and nonwoven felts are available in a number of materials, including polypropylene, polyester, acrylic, modacrylic, nylon, polyethylene, Saran®, Teflon, PVC and glass.

Much could be written about the fine points of available fabric filter cloths; suffice it to say that each of the various materials must be considered within the parameters of cost, available permeabilities, temperature limitations, chemical compatibility, mechanical strength, dimensional stability, abrasion resistance, cake release characteristics and blinding tendencies.

Economics is the main reason that most tubular pressure filters utilize elements covered with fabric cloth media. Even the more sophisticated synthetic filter fabrics are usually less expensive than woven wire cloth and porous media.

Furthermore, with porous media, wedge wire or wire cloth elements, the septum either *is* the entire element or is built into the element and is a part of it, so that replacement of the septum itself really means replacement of the entire element. Fabric sleeves (either seamless or sewn) on tubular elements are generally clamped or tied on the element support and backup structure, can be readily removed for laundering, soaking in acid or caustic solutions, etc., and reused until they tear or become too severely blinded. When these sleeves can no longer be reused, it is relatively easy and inexpensive to replace them with new ones.

Although it is always wise in selecting backwashing filters to try to avoid misapplying a design that tends to blind rapidly and resist cleaning by backwashing, the consequences in terms of labor and replacement media cost are not nearly as serious when cloth sleeves become blinded as when permanent, costly septa exhibit severe blinding tendencies.

Tubular Element Reverse Hydraulic Shock Filters (Figure 5.16)

The term "backwashing" describes a filter element cleaning method that consists basically of reversing flow of liquid (usually water) through a filter with the intent of removing all accumulated solids from the elements and restoring the filter elements to their original clean condition. However, in actual practice, this ideal is never achieved, and the whole problem of blinding of the filter media or septa is relative, based on the type of media or septa, the nature of the solids being retained, whether filteraid precoats are being used, and many other factors. It should be clear then that blinding is a major problem in backwashing-type tubular filters, as well as in other types of filters, and, therefore, any filter design that minimizes blinding tendencies certainly offers considerable advantage over most other types.

The hydraulic shocking type of tubular filter is one that tends to do a better job than most others of keeping filter media or septa from blinding. There are several designs available today, but one extensively used design is essentially a vertical pressure chamber inside of which are suspended from a tube sheet or header-lateral at the top of the chamber a series of molded polypropylene tubes covered with seamless flexible fabric sleeves, usually polypropylene but available in other fabrics. The bottom of the chamber is conical and the top of the chamber is equipped with a large air dome. The inlet is at the bottom and the outlet is at the top above the tube sheet.

Like most pressure filters, the operation of this type filter is cyclical. Unfiltered liquid enters the chamber at the bottom, flows through the tubes, depositing solids on the sleeves which have either been precoated or not, depending on the nature of the solids and the clarity requirements. The filtrate flows upward through the tubes into the head and out through the cover. To clean the filter, the filtrate discharge valve is closed while unfiltered liquid continues to enter the chamber. As a result, filtrate builds up in the pressure dome, at the same time compressing the air above it. When this pressure reaches 50–75 psig (3.52–5.3 kg/cm^2), depending on the pump shutoff pressure capability, a quick-opening drain valve in the bottom of the

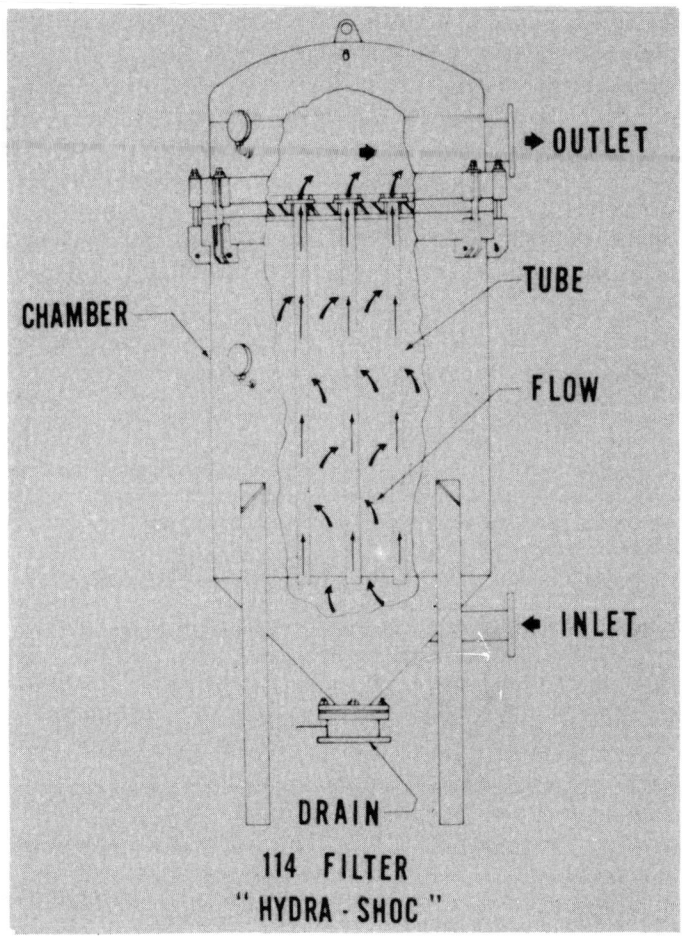

Figure 5.16. Diagram cutaway of tubular element reverse hydraulic shock filter. (Courtesy Industrial Filter & Pump Manufacturing Company.)

chamber is opened, thereby allowing the liquid in the filter to escape and the compressed air in the dome to expand, pushing filtrate ahead of it. Momentary flowrates during reverse flow of the filtrate are tremendously high. Actual cleaning of the tubular surfaces is a two-step process, although the time sequence between the two steps is only a fraction of a second. Step 1 is an actual expansion of the seamless fabric sleeves that cover the rigid plastic tubes. This causes the cake to crack along the entire length of the tubes. Step 2 is a reverse hydraulic flow at extremely high rates, which dislodges the cracked filter cake from the tubular surfaces. Cake is discharged in a slurried state from the bottom of the filter chamber.

This type of filter is used primarily as a precoat filter, and in many instances, a filter cake of up to ¾ in. (1.9 cm) thickness is built up on the tubes before the hydraulic shock cleaning takes place. However, it can also operate on a very short cycle basis, with frequent removal of a relatively thin cake to maintain high average flowrates per square foot of filter area.

Units of this type are available in the range of approximately 50–600 ft^2 (4.6–56 m^2) of filter area. In addition to fabric sleeves, the tubes may be made of wire mesh, porous media or wound wedge wire. However, these rigid media do not offer the advantage of the snap-out, stretching and self-cleaning capabilities of the fabric sleeves.

In general, there are several advantages to backwashing and reverse hydraulic shock-type tubular filters, as long as the one major disadvantage of wet (reslurried) solids discharge can be tolerated:

1. Tubular elements often give higher filtration rates per unit area than flat elements. This is true particularly in filteraid applications where a filter cake builds up on the elements, because the surface area increases with cake thickness and this increasing filter area tends to counteract the growing cake resistance, which, of course, tends to reduce filtration rates.
2. Off-stream time for cleaning is less than for most filters. Even large-capacity precoat filters can be completely cleaned and put back on stream in 15–30 minutes.
3. These filters are available in a wide variety of construction materials and can be applied to even the most corrosive and toxic liquids.
4. Because of the relatively high rates of flow per unit of filter area, the initial cost per gpm(1/min) of capacity is very often less than many other types of filters.
5. They are relatively low in maintenance labor and parts if properly applied and operated.
6. Because of their simplicity, they are generally more easily and inexpensively automated than most precoat filters.

Flexible-Tube Filters

Flexible-tube filters are a somewhat unusual adaptation of standard tubular backwash filters, in that the filter tubes, approximately ¼–¾ in. (0.63–1.91 cm) in diameter, are made of a flexible stainless steel or fabric braid, which its manufacturer claims to greatly reduce the blinding tendencies of most rigid tubular septa.

The braided septa must be precoated heavily with a filteraid (Figure 5.17). Contaminants are accumulated on the precoat until flowrate or pressure drop indicate that the precoat has become blinded and then, either manually or automatically, a procedure called "bumping" takes place, in which the whole filter element assembly is rapidly moved up and down several times so that the tubes flex and twist and purge themselves of the blinded filter cake. In the process, dirt and filteraid are mixed together, and when the elements are cleaned, the solids are reapplied to the braided tubes by a short recirculation period and the filter put back on stream.

This procedure can, of course, be carried out only a certain number of times

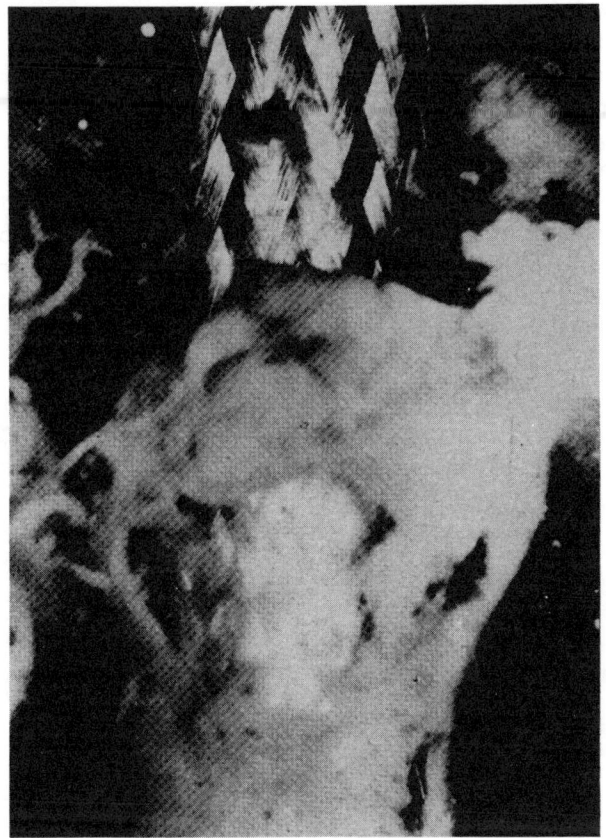

Figure 5.17 Close-up action photograph showing "bumping" procedure used in cleaning braid filter septa. (Courtesy Chem-San International Corp.)

before the proportion of contaminant to filteraid is such that filtration rates diminish, and then the accumulated solids are discharged from the filter and replaced with a fresh precoat charge.

These filters are widely used in the swimming pool and dry cleaning markets, and also have many general industrial applications. Their main advantages are the reduced tendency toward filter septa blinding and economical filteraid usage.

Centrifugal Discharge Filters (Figure 5.18)

Centrifugal discharge filters could be classified under horizontal plate filters because they are, in fact, vertical-tank, horizontal-plate filters, which offer the advantages of horizontal plate filtration but which are cleaned by spinning the cartridge assembly.

In a typical centrifugal discharge filter, the internal cartridge consists of a number of round filter discs or leaves mounted on a vertical tubular shaft, which can be rotated by an external motor. The discs or leaves are separated by adjustable spacers, depending on cake thickness requirements. Used normally with filteraids, the septa are wire mesh or woven fabric.

Precoating, filtration and cake washing, if required, are performed in a manner similar to most other filters. However, when the cake is ready for discharging the cartridge is rotated by the motor, and simultaneously water or other liquid is pumped backwards through the shaft and the leaves, and the reslurried solids are discharged from the filter. The total cleaning time is claimed to be about one minute.

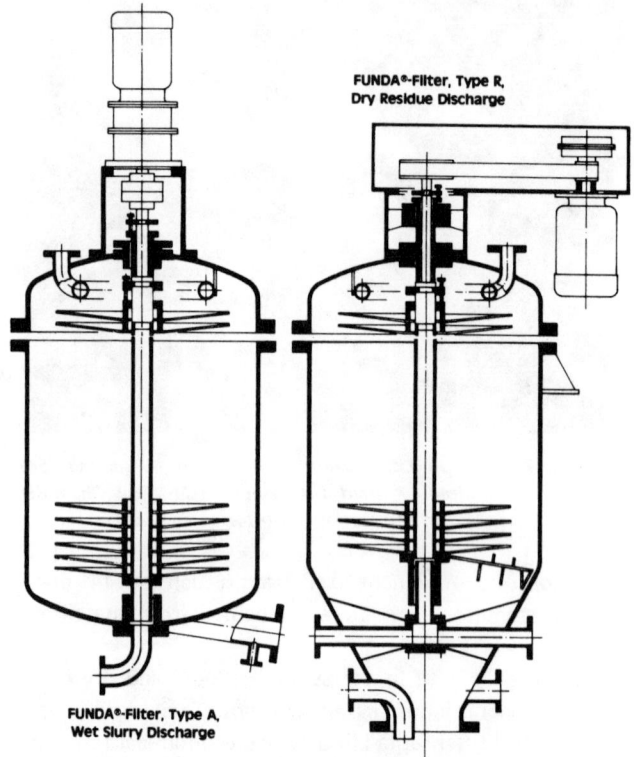

Figure 5.18 Diagram of typical wet and dry cake discharge centrifugal discharge filter. (Courtesy Chemapec Inc.)

Construction materials are generally stainless steels and other corrosion-resistant materials, and lined tanks are available. Filter areas are in the range of 20 ft² (1.9 m²) up to approximately 3000 ft² (280 m²).

Advantages claimed for these filters, in addition to the previously discussed advantages of other horizontal cake filters, are:

1. low operating labor and downtime;
2. designed for cleaning without frequent opening;
3. no unfiltered liquid heel; standard scavenger plates;
4. ready adaptability to fully automatic operation; and
5. dry cake discharge possible in some models.

Disadvantages are high vibration forces during cleaning, high cost per unit area, high maintenance usually connected with moving parts, seals, etc., difficulty of controlling blinding of filter septa, and difficulty of removing and cleaning blinded septa.

DRY CAKE DISCHARGE FILTERS

Although dry cake discharge filters are largely used with filteraids, either as precoat only or with body feed, there are many applications in which the solids to be removed either do not require the use of filteraids, or filteraids cannot be used, either because the separated solids are of some value and cannot be mixed with a foreign material, or because pickup of silica or some other minor impurity in the filteraid is unacceptable in the filtrate.

However, the operation, cleaning and advantages of dry cake discharge filters are the same in either case.

Plate-and-Frame Filter Presses (Figure 5.19)

The plate-and-frame filter press is, without question, the most widely used and most familiar dry-cake-discharge pressure filter in existence. Its flexibility, ruggedness, simplicity of operation, relatively low cost in wood, polypropylene and in steel-and-cast-iron construction are hard to match with any other pressure filter.

Essentially, it consists of a horizontal floor-mounted chassis or skeleton consisting of two end supports, connected by two side rails. At one end is a fixed head and on the other a movable head. Between the two heads and supported by the side rails are square filter plates and frames mounted on all or part of the chassis, forming a number of filter chambers. These chambers may be formed either by alternate plates and frames or by adjoining recessed plates only.

The plates are made with drainage channels resulting from closely spaced raised pyramids, parallel grooves or radial grooves, and the frames are hollow and provide space for filter cake accumulation. Filter cloths are placed between the plates and frames (or between recessed plates), the chambers are closed and tightened by a heavy screw or hydraulic ram which holds the plates and frames together under considerable pressure, and the cloths serve both as filter septa and as gaskets.

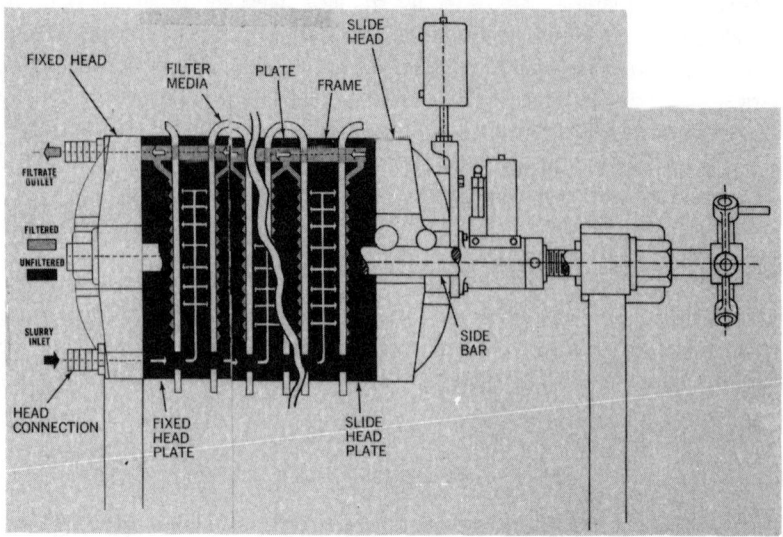

Figure 5.19 Sectional view of a fully assembled corner feed, closed discharge filter press, showing plates, filter cloth and frames in place on filter press skeleton and indicating flow of material. (Courtesy D. R. Sperry & Co.)

Conventional plate-and-frame presses are available in filter areas from a few square feet with 6 in. (15.2 cm) square plates and frames up to several thousand square feet (several hundred square meters) with 56 in. (142 cm) square plates.

In addition to plate-and-frame versus recessed-plate configurations, innumerable plate configurations are available, involving variations of feed and discharge channels, frame thicknesses, cake washing ports, open versus closed discharge, etc., but it is not within the scope of this chapter to discuss all these design variations.

More recent innovations of filter press design (Figure 5.20) have provided gasketed plates to reduce leakage, metallic as well as textile filter septa, caulked-in filter cloths and much larger sizes, up to 64 in. (163 cm) square plates and filter areas up to 5,000 ft^2 (465 m^2). These large units are made practical by mechanical plate shifters and relatively labor-free discharging of filter cake.

Plate-and-frame presses are available in many construction materials as well as lined and coated constructions. Filter septa, usually fabrics, are most often cotton or one of the standard synthetic filter fabric septa already discussed.

Their advantages are legion. Besides flexibility, ruggedness and simplicity, they normally provide a cake dry enough to handle as a solid waste (as long as the frames or recessed plates are filled); they have very little unfiltered liquid "heel" at the end of the cycle; they are not easily damaged by trying to overload them with solids; and the filter cake can be readily washed with water or solvents to recover desirable materials.

Figure 5.20 Large recessed plate filter press used in a sludge dewatering plant. Cake capacity is 18 yd³ (4.6 m³) of dry sludge. (Courtesy Passavant Corporation.)

However, conventional-design filter presses are generally high in cleaning labor and downtime; the maintenance and replacement of filter cloths is often an appreciable expense; they may leak rather badly, since most cloths do not gasket well, so the housekeeping is usually poor; and they are quite expensive in corrosion-resistant materials such as stainless steel.

Some of these disadvantages have been minimized by the newer plate-and-frame designs (Figure 5.21), but these are generally economical only in the larger sizes.

Pressure-Leaf Filters

The configurations of both vertical- and horizontal-tank pressure-leaf filters have already been discussed earlier, since both types can be adapted to either wet or dry cake discharge (Figure 5.22). However, there are so many aspects of superiority of the horizontal tank filter over the vertical for dry cake discharge that relatively few vertical-tank units are used in this way. The inability to inspect the leaves effectively, the possible hang-up of solids on a bottom manifold*, and the distance

---

*It is only fair to say that this can also be a problem in horizontal-tank designs with bottom outlet manifolds, but at least the problem can be readily seen and dealt with.

Figure 5.21 Modern mechanized filter press with plate shifter. (Courtesy D. R. Sperry & Co.)

from the source of vibration to the bottoms of the leaves are just three reasons why the vertical tank design may be more troublesome in dry-cake-discharge applications.

Dry-cake-discharge, horizontal-tank filters are essentially the same design as those discussed earlier, but the design most widely used is the fixed-leaf bottom outlet. Rotating leaves with central manifolds are generally not yet extensively used for dry cake applications.

The main design difference between wet- and dry-cake-discharge filters is the use of vibrators instead of sluicing devices. A pneumatic vibrator or vibrators are generally mounted externally on the cover, and the shaft attached to the piston enters the tank through a gland built into the cover. The shaft is attached to a vibrator bar, which normally is bolted to the tops or the sides of the leaves (Figure 5.23).

Normal cake discharge is handled in this way: when a filtration cycle is completed, the filter heel is blown out through the drain valve by compressed air brought in at the top of the filter (often the outlet manifold valve must be partially open to prevent dropping of the filter cake). As soon as the tank is empty, the drain valve is closed and the outlet manifold fully opened. Generally, by the time compressed air is blown through the cake at a pressure of 5–15 psig (0.35–1.05 kg/cm^2) for 10–15 minutes, the cake is sufficiently dry to be friable (moisture

Figure 5.22 Vertical-tank leaf filter designed for dry cake discharge. Note bottom drop door and bag shaker device on cover. (Courtesy Industrial Filter & Pump Manufacturing Company).

content 40–70%). The air is then shut off, the filter opened, the leaf carriage rolled out, and the vibrator or vibrators actuated for whatever number of minutes are required to remove the cake into a hopper or conveyor installed below the carriage.

Rarely does a vibrator remove filter cake 100%, but once the filter is 90–95% clean, it can generally be closed up and a new filtration cycle begun. Off-stream time is generally in the range of 30–60 minutes.

Before attempting to apply a dry-cake-discharge filter of this type to a given service, it is well to have had some assurance by either laboratory filtration testing or actual pilot-scale operation that a filter cake of at least 3/8–1/2 in. (0.95–1.27 cm) thickness can be accumulated within reasonable flow and pressure operating

Figure 5.23 Dry cake discharge leaf filter showing double pneumatic vibrators and vibrator bars attached to all-metal leaves. (Courtesy Industrial Filter & Pump Manufacturing Company.)

ranges, either with or without filteraids. Otherwise, the air-blown filter cake may be too light in weight to separate readily and fall from the leaves under vibration, and must then be removed by manual scraping. Also, it is desirable in *most* cases that an adequate filteraid precoat be applied to ensure a clean separation at the interface between the filter cake and the septum.

A variation of the standard dry-cake-discharge horizontal leaf filter design is offered by one manufacturer to provide mechanical compression of the filter cake both to compact it in place and to provide a dryer cake than can be obtained with air blowing alone (Figure 5.24).

In this design, impermeable plasticized diaphragms are positioned between the filter leaves, which can be either wire cloth or fabric-covered, as shown in Figure 5.24. There are two closely spaced diaphragms between each pair of leaves, and, when the cake has been built up on each leaf to the diaphragms, flow through the cake stops and full shutoff pump pressure (50–75 psig (3.5–5.3 kg/cm^2) is normally needed) is applied to the backs of the diaphragms, thereby compressing the filter cake and removing moisture sometimes to as low as 25 or 30%.

Leaf spacing and diaphragm positioning is often such that a filter cake of 2–3 in.

Figure 5.24 Horizontal leaf filter with compression diaphragms between leaves for maximum filter cake drying prior to vibrator discharge. (Courtesy Industrial Filter & Pump Manufacturing Company.)

(5.1—7.6 cm) can be built up on the leaves before diaphragm contact and compression are achieved, but laboratory test work on a specific material is desirable to determine cake thickness feasibility and whether the physical compression provides the cake dryness desired.

For large-capacity process fluid filtrations, the choice for dry cake discharge applications if often between plate-and-frame and horizontal-tank, pressure-leaf filters.

The advantages previously attributed to filter presses are corresponding disadvantages of leaf filters; that is, leaf filters are a little more complicated in their operation and require slightly more skilled operators; there is a liquid heel at the end of each cycle to handle in some way; leaves may be damaged by overloading with solids, by rough handling or improper vibration; and leaf filters are not ideally suited to cake washing. However, several major advantages have been responsible for the increasingly widespread use of these filters in the processing industries.

First, cleaning labor and offstream time are generally far less for horizontal-tank leaf filters. Second, flowrates per unit area are often appreciably greater with leaf filters, particularly when comparing wire-cloth-covered leaves with heavy, low-permeability filter press cloths. Third, the filter station is normally far cleaner because of the lack of product leakage during filtration and the method of cake removal. Fourth, filter leaves require little handling for maintenance and repair, as compared to the continual handling, laundering, maintenance and relatively frequent replacement of filter press cloths. Fifth, where corrosion-resistant construction is required for fiarly large filters, relatively lightweight leaf filters are often a good deal less expensive than filter presses with their heavy castings.

Once again, it should be said that many of the disadvantages of the plate-and-frame filter press have been overcome or greatly reduced by the newer designs involving gasketed plates, mechanical plate shifters, compressed air cake removal and automated operation with microcomputers. In any given size range, therefore, a thorough evaluation of economic and operational features, as well as capital outlay versus operating costs, must be made.

Horizontal Plate Filters (Figure 5.25)

A third type of pressure filter that has been used for many years in dry-cake-discharge applications is the horizontal plate design. This is a vertical-tank filter, usually offered in 20 in. (50.8 cm) and 36 in (91.4 cm) diameters, with an internal cartridge made up of a series of horizontal plates held together as an assembly in such a way that the entire cartridge assembly can be removed as a unit from the filter for external disassembly and cleaning.

The cartridge construction most commonly used is the type shown in Figure 5.25. It consists of alternate plates, which support the filter cake, and spacer rings, which are usually 3/8—2 in. (0.95—5.1 cm) thick, and which separate the filter plates to allow for filter cake space.

Filter paper discs are usually used as the media or septa. Generally filteraid is used with these filters, but occasionally they are used as trap filters, with particle removal capabilities only as low as the grade of filter paper used. (Although filter paper manufacturers do not rate their grades by micrometer, nominal ratings can be considered to be roughly from 5—10 μm upward).

These filters are available with what is called a scavenger plate in the bottom of the filter, which allows the unfiltered liquid heel to be filtered by air or gas pressure applied to the top of the vessel.

There are two basic types of horizontal plate filters. In one type, the liquid flow is into the filter tank, through holes in the spacer rings, then through the paper media, through holes in the inner diameter of the filter plates to the outlet in the vertical center shaft. In the other design, the liquid flow is exactly the reverse. Liquid enters through a center vertical manifold, is distributed outward over the media, flows through the media and out through holes in the outer diameter of the plates and is discharged from the bottom of the filter tank (Figure 5.26).

Figure 5.25 Horizontal plate filter with cutaway view of plate assembly. (Courtesy Sparkler Manufacturing Company.)

The advantage claimed for the latter type is that where the recovery of all liquid in an entire batch is essential, the small amount of unfiltered liquid remaining in the inlet piping, the central inlet manifold and above the cake can be forced through the entire filter area by air or gas pressure in a relatively short time. In the former, the unfiltered liquid in the tank must be forced through the relatively small area of a separate scavenger plate, and, in some cases, this may require a considerable amount of time, possibly longer than the time required for the filtration of the entire batch.

At the end of a batch or a filtration cycle, signaled by unacceptably low flowrate or high pressure drop, the filter is usually blown with compressed air or gas to dry the cake, the filter is opened and the cartridge assembly removed by means of a chain hoist and set on the floor. The cartridge is then disassembled, plate by plate, with filter paper discs and accumulated cake discarded to waste, the plates assembled and redressed with fresh filter paper and hoisted back into the filter.

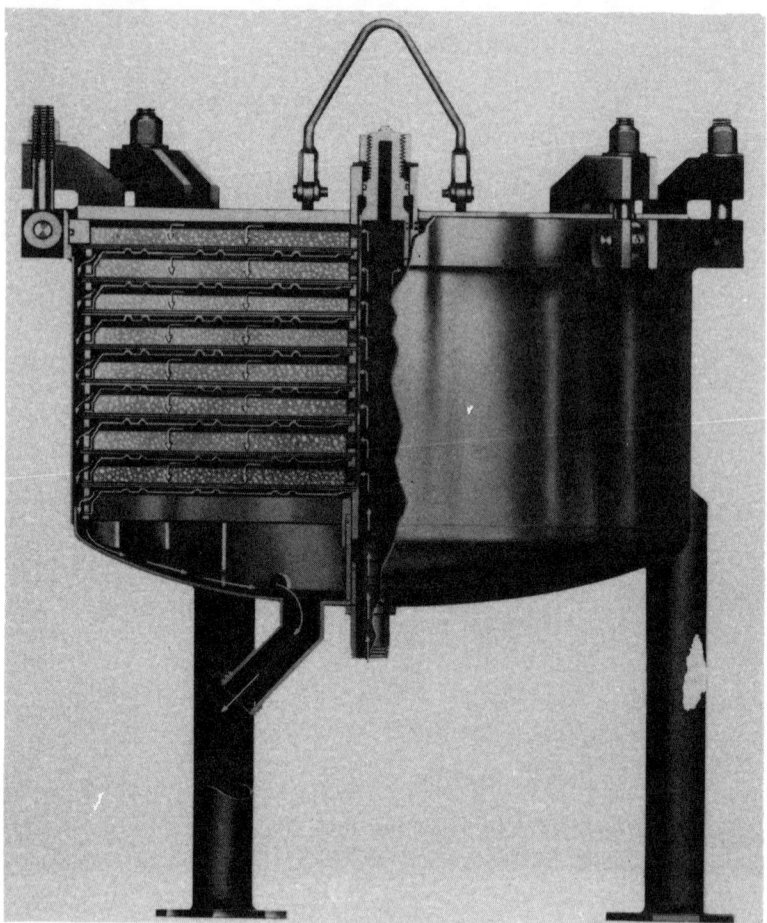

Figure 5.26 Horizontal plate filter with cutaway view showing flow from center inlet outwards. (Courtesy Niagara Filters Division, Ametek, Inc.)

There are two variations of the horizontal plate design in which both sides of the filter plates are used. In one, the plates are covered with cloth or wire mesh and are simply washed off externally and reused without disassembly of the cartridge. In another, filter paper is used over perforated metal plates. However, where wet cake discharge is acceptable, there are so many types of filters that can be cleaned more quickly and easily that these variations of the horizontal plate filter are seldom used.

Horizontal plate filters are generally available in areas from 3 ft^2 (0.28 m^2) up to about 200 ft^2 (18.6 m^2) with cake-holding capacity up to 20 ft^3 (0.57 m^3) and in all normal metals and with lined tanks and molded plastic plates for highly corrosive applications.

As already mentioned, media or septa on the plates are normally one of many available grades of filter paper made by at least two American industrial filter paper manufacturers to fit all standard sizes. A type of nonwoven disposable filter fabric is also sometimes used. Infrequently, reusable fabric and wire cloth are used as septa.

The horizontal plate filter provides a number of advantages:

1. It has the ability to completely filter a batch without any liquid heel.
2. Horizontal cake position provides greater stability, particularly for intermittent operation, in which vertical cakes would be more seriously disturbed.
3. Use of inexpensive disposable filter paper septa often permits dispensing with the precoating step by simply recirculating the feed liquor until suitable clarity is achieved.
4. Since the septa are meant to be discarded, the septa blinding problems common to many filters do not exist.
5. Filter elements can be readily disassembled for inspection and thorough cleaning.
6. Filteraid bleedthrough, which can be a problem with the septa used in many filters, is greatly reduced if the proper filter paper is used. However, if plates are warped and/or dented, even use of tight, heavy filter paper will not prevent bleedthrough.
7. Simplicity of assembly and operation make it almost foolproof.

Because of this last advantage, horizontal plate filters are often used as final polishing filters or combination polishing/trap filters.

There are three disadvantages of horizontal plate filters:

1. They require manual labor and handling for cleaning, on the order of ½—1½ man hours per filter cycle. This also makes them undesirable for highly acid and caustic applications, high-temperature applications, and applications involving toxic and highly volatile liquids. Off-stream time can, of course, be reduced to minutes by having an extra cartridge assembly ready to exchange for a dirty one at the end of each cycle.
2. Since they are available in sizes only up to about 200 ft^2 (18.6 m^2), high-capacity applications require multiple units, which means higher capital outlay and even greater cleaning labor.
3. Overhead suspension equipment is required for cartridge removal.

Centrifugal Discharge Filters

Centrifugal discharge filters, like pressure-leaf filters, are as readily adaptable to dry-cake discharge as to wet-cake discharge.

After heel filtration, cake washing (if desired), and drying by air or steam blowing, the cake can be discharged by spinning the cartridge without using any backwashing fluid through the filter leaves (Figure 5.18).

Horizontal Tray Filters

Horizontal tray filters are claimed to combine the advantages of horizontal cake stability with either automatic or low-labor dry-cake discharge. There are several designs available (Figure 5.27).

One design offers a horizontal tank with filter plates or trays mounted length-

Chapter 5. Alfred Trumpler

SPARKLER FILTER MODEL HRC
A UNIQUE SPARKLER CONCEPT

The Model HRC combines the advantage of horizontal plate cake stability with automatic cake discharge, an innovation engineered by Sparkler. It is available in capacities up to 300 sq. ft. of filtering surface.

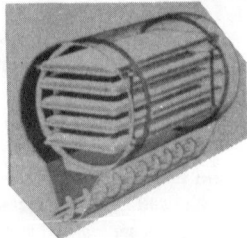

Plates in horizontal position during precoat, filtration, cake washing, and drying cycles.

Plates in vertical position during the cleaning cycle.

Figure 5.27 Left view shows plates or trays in horizontal position for precoating, filtration, cake washing and drying. Right view shows plates in vertical position for cleaning.

wise in a horizontal position during precoating, filtering, cake washing and/or air blowing, then to discharge the cake, the plate or tray assembly is rotated inside or outside the tank to a vertical position and the dry cake discharged, either into an external hopper or into a screw conveyor built into the bottom of the filter tank. Another design also offers external handling of each tray for either dry or reslurried cake discharge. Both designs offer individual tray filtrate discharge.

Basically designed as precoat filters for cake thicknesses up to 2¼ in (5.7 cm), the trays or plates are generally constructed with wire mesh or synthetic woven fabric filter septa, but can also be adapted to the use of filter paper.

Sizes range from about 10 ft^2 (0.93 m^2) up to about 300 ft^2 (27.9 m^2) of filter area, but they are used primarily in sizes of 50 ft^2 (4.6 m^2) or less.

Although not too many of these filters have found their way into the processing industries, possibly due to their intermediate size range and relatively high cost per unit area, they offer some important advantages where properly applied:

1. ideal for cake washing and air drying;
2. horizontal cake stability (as discussed in the previous section);
3. little or no unfiltered liquid heel; and
4. internal self-cleaning models ideal for handling toxic, volatile materials.

"Inside-Out" Tubular Filters (Figure 5.28)

All tubular filters discussed so far are designed for liquid flow from outside of

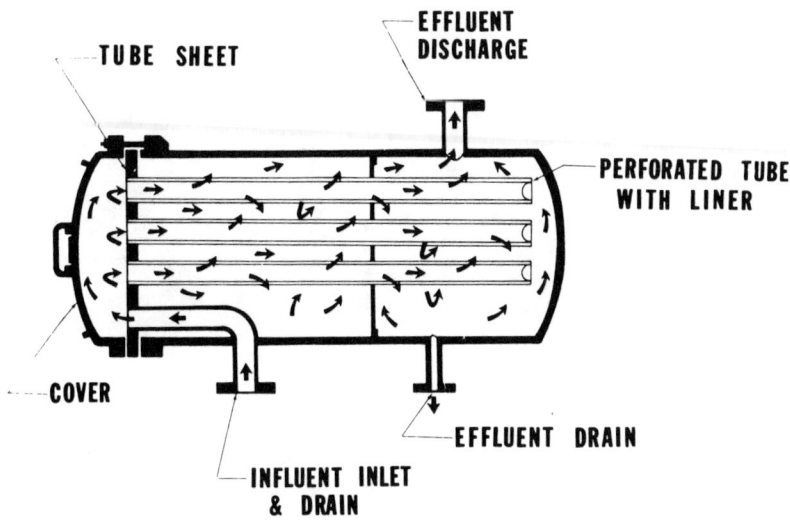

Figure 5.28 Diagram of tubular filter with inside-out liquid flow. (Courtesy Industrial Filter & Pump Manufacturing Company.)

the tubular elements toward the inside, but one type is designed for inside-to-outside flow. This filter consists of one or more perforated metal tubes (usually stainless steel), which are mounted either horizontally or vertically and in which are inserted rolled-up sheets of filter paper or similar expendable media. Used either with or without filteraid, generally as a final trap or polishing filter, the removed solids accumulate inside the tubes; cleaning is accomplished by manually removing the dirty filter paper liners and discarding them as dry waste. It is available from 2 ft^2 (0.2 m^2) to approximately 200 ft^2 (18.6 m^2).

Advantages are that (1) by pressurizing the vessel with air, virtually all the unfiltered liquid can be filtered so the heel is almost nothing, (2) filter paper media liners are very inexpensive, and (3) it can be used satisfactorily for intermittent operation. The main disadvantages are that it provides limited cake capacity and requires relatively high manual labor for cleaning and long offstream time between cycles.

FILTERS DESIGNED TO OPERATE NORMALLY WITHOUT FILTERAIDS

Cartridge Filters

Cartridge filters (Figure 5.29) are probably the most widely used of all pressure filters on the market. They are made in so many different configurations and with so many minor variations by their many manufacturers that it would be difficult in this chapter to adequately describe and evaluate them all.

Chapter 5. Alfred Trumpler

Figure 5.29 Variety of filter cartridges offered by one manufacturer. (Courtesy The Hilliard Corporation.)

Although it is not always a completely clear distinction, every cartridge can generally be classified as offering either primarily depth or primarily surface filtration, and can also be classified in terms of cleanable (reusable) or throwaway (expendable).

Relative to the other parameters, which are referred to in the "Introduction," cartridges are practically all used in low solids applications, and although there are some situations in which filteraids are used with cartridges, by far the majority are used without filteraids.

As far as nominal versus absolute micrometer ratings are concerned, this section will deal only with the nominal category, since probably 99% of the cartridges used are in this category, simply because the price premium required to produce absolute cartridges is not economically justifiable in 99% of the industrial cartridge filter applications. Absolute cartridges will be discussed later in this chapter.

Throwaway (Expendable) Cartridges — Certainly the largest portion of the throwaway cartridge market is divided among depth-wound fiber cartridges, resin-impregnated or molded fiber cartridges, and pleated paper cartridges. The wound and molded configurations are considered primarily as depth filters, and pleated paper primarily as surface filters.

Wound cartridges (Figure 5.30) are manufactured by winding roving of cotton, rayon, polypropylene, glass, acrylic, modacrylic, polyethylene, polyester, nylon, cellulose acetate and even Teflon, on a plastic or metal core. For many years, these cartridges were available in standard diameters of 2½–3 in. (6.4–7.6 cm), and in lengths from 4–10 in. (10.2–25.4 cm). In high-capacity housings, multiples of 10-in. (25.4 cm) cartridges were used with or without spacers between the cartridges. Now wound cartridges are available in lengths up to at least 50 in. (127 cm).

By the nature of the rapid back-and-forth motion of the high-speed winding machines, the diamond-shaped patterns of the fiber roving are quite small with a rather tight winding at the core, and as the diameter of the winding increases, the

Figure 5.30 Three different nominal micrometer-rated 10 in. fiber-wound filter tubes. (Courtesy Filterite Division Brunswick Corporation.)

diamond pattern enlarges and openings between the roving increase, with the result that the outer winding is considerably more permeable than the inner.

This variation of winding permeability of wound cartridges between inside and outside gives us what is called depth filtration: larger particles are stopped near the outside of the cartridge and finer particles may well proceed through a tortuous path in the winding before being stopped near the center core. Therefore, particles are generally retained throughout the entire depth of the winding, and since the solids are, to a large extent, separated from each other and do not form an impermeable layer, the ability of the depth-wound cartridge to hold a relatively large quantity of solids before becoming blinded is one of its good features.

Not all depth-wound cartridges are made in the same way. For example, one major variation of the standard winding process described above is that of blowing loose fibers into the cartridge during the winding process so that there are in effect thin blankets of fiber between relatively open layers of wound roving, and actual filtration is performed by these externally introduced fiber layers rather than by the wound roving itself. Nominal micrometer range for wound filter tubes (cartridges) is ½ to approximately 350).

The "molded" (Figure 5.31) cartridge is so named because of its solid, molded appearance and feel, but few of the so-called molded cartridges are actually molded.

Chapter 5. Alfred Trumpler

Figure 5.31 Typical grooved molded cartridge. (Courtesy AMF Cuno Division.)

Figure 5.32 Another type of "molded" cartridge, which is actually a resin-impregnated wound cartridge. (Courtesy Filterite Division, Brunswick Corporation.)

The one thing they all have in common is that they are impregnated with a phenolic or melamine resin and subsequently cured, which gives them dimensional stability.

Otherwise, they are made of a variety of materials, such as cellulose, wool, acrylic, viscose rayon, polyester and glass fibers. Some are formed initially on vacuum tubes, others are formed into a sort of a cylindrical bat, still others are wound roving (Figure 5.32). Some have perforated center cores; some have no cores. Whatever the process, all are depth media, with the fibers, of whatever length they may be, locked into position by the cured resin. All claim to have graded density, or variable permeability, with the outer portion considerably more permeable than the inner portion. Some are grooved to provide additional surface area.

Nominally rated micrometer range of most resin-impregnated cartridges is 5–125.

Pleated paper cartridges (Figure 5.33) differ from wound and molded cartridges in that they are primarily surface media, and since surface media have a much greater tendency to blind with a relatively small amount of compressible and slimy solids, large surface area is obtained by getting as many pleats as possible into a cartridge of a given overall size.

Most pleated cartridges are made simply by forming the desired size piece of pleated paper into a cylindrical shape, sealing the joint, then securing and sealing this pleated cylinder into plastic or metal end caps with some type of cement, such as epoxy. The assembly may or may not be strengthened or rigidized by a center plastic or metal perforated core.

The term "paper" is used loosely in describing these cartridges but, in many cases, the filter medium is not really paper, but a paper-like medium made with glass fiber, nonwoven synthetic fabrics, spun-bonded materials, etc. In most cases, these media are impregnated with various resins which actually are not cured until after the media have been pleated.

Although we referred to these cartridges as being primarily surface filters, a form of depth filtration is provided in many of the more costly pleated cartridges by using two or three layers, usually of different media. In this situation, the filter medium itself is followed by a downstream trap filter medium, often a relatively migration-free material, and if a third layer is used, this is generally an inexpensive prefilter, which serves to pick up large particles and, at the same time, to protect the filter medium from physical damage. Nominal micrometer levels of pleated cartridges are from 0.1–30.

Although most cartridge manufacturers produce single-cartridge (Figure 5.34) or multiple-cartridge housing designed to properly hold and seal their own cartridges, there is a degree of standardization among both housings and cartridges so that interchangeability of cartridges within a certain sizing system is common. Although there is a sizing system in the industrial oil and hydraulic industry, which is built around pleated paper cartridges approximately 7×18 in (17.8 $\times$ 45.7 cm), the system used almost completely in the processing industries is based on a unit cartridge size of 1–1-1/16 in (2.5 to 2.7 cm) i.d., 2½–2¾ in (6.4–7.0 cm) o.d. $\times$ 9¾–10 in (24.8–25.4 cm) length.

Certain cartridges are also made in lenghts as low as 4 in. (10.2 cm) and many are now offered in multiples of 9¾ or 10 in. (24.8 or 25.4 cm), up to 5 standard lengths or approximately 50 in (127 cm).

The important thing about this interchangeability is that the user is generally not

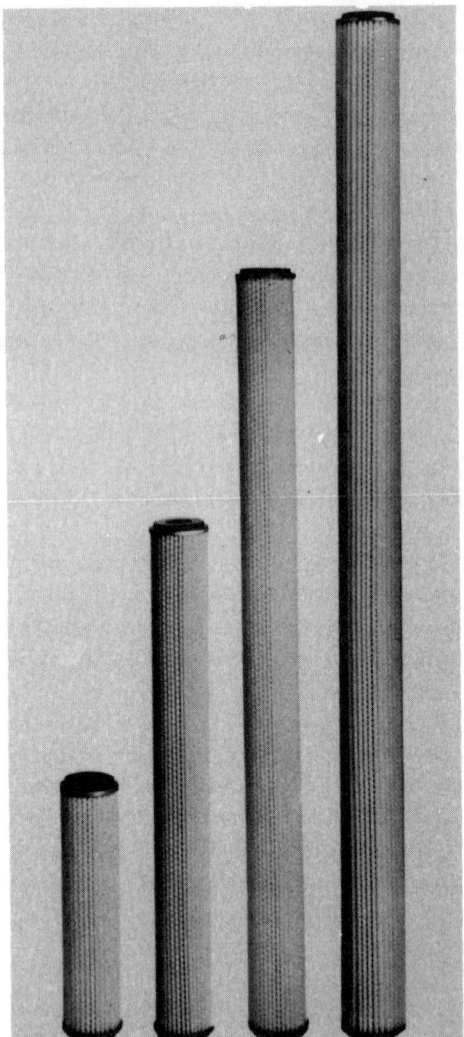

Figure 5.33 Pleated paper cartridge available in lengths from 10–40 in. (Courtesy Filterite Division, Brunswick Corporation.)

locked into only one supplier of cartridges when he purchases a certain housing. Therefore, he can evaluate various types of cartridges on the basis of his particular needs and, except for advance considerations of chemical compatability, leachable impurities, temperature limitations and migration characteristics, it is nearly impossible to prejudge which type of cartridge will be the most effective and economical for a specific application.

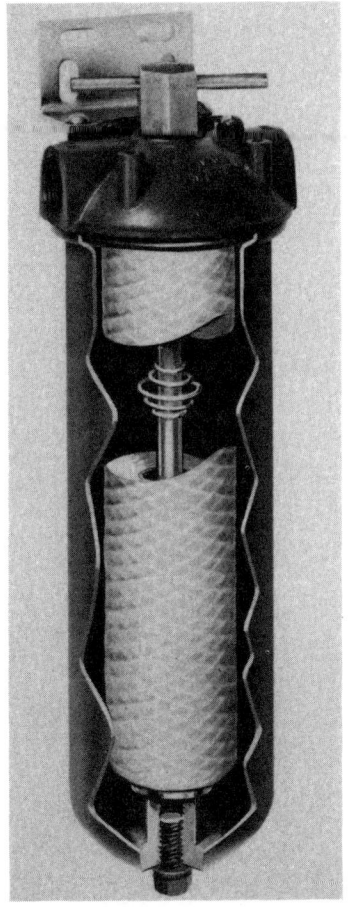

Figure 5.34 Typical slingle-cartridge housing, showing cutaway of cartridge. (Courtesy Filterite Division, Brunswick Corporation.)

A knowledgeable and experienced technical filtration person may be able to guide the user in certain directions and indicate some probabilities, but without actual testing in service, no one can be dogmatic as to whether a wound, molded or pleated paper cartridge will provide the desired degree of contaminant removal at the lowest cost per gallon filtered.

Users must be cautioned when interchanging cartridges, however, that different manufacturers have different methods for sealing the ends of their cartridges, and it must be made certain that proper end sealing exists when using one make cartridge in another manufacturer's housing.

There is yet an additional large variety of nominally rated throwaway cartridges available to the process designer and engineer. There are felt cartridges, cartridges with replaceable polypropylene felt sleeves (Figure 5.35) cartridges made of stacked cellulose discs, microporous plastic media, Fuller's earth, activated carbon (Figure 5.36), glass microfiber (Figure 5.37), molded viscose fibers and many

Chapter 5. Alfred Trumpler

Figure 5.35 Filter housing cutaway showing a low-cost polypropylene felt sock cartridge in place. These felt cartridges are available from 2–200 μm nominal. (Courtesy Technical Fabricators, Inc.)

others. Each type has certain features and application areas of particular strength, and maintains a certain niche in the vast throwaway cartridge market.

Cleanable (Reusable) Cartridges — Many of the filter cartridges commonly cleaned and reused were discussed earlier under Tubular Element Filters, p. 123. In that section, the various materials, such as porous ceramic and stainless steel, wire mesh, fabrics and wound wedge wire were considered as filter septa, to be used with filteraids. In this section, these materials in cartridge form are to be considered as filter media, and, therefore, the problems of cleaning and reuse are more critical.

Porous ceramic cartridges (Figure 5.38) are available in micrometer ranges from 5 upwards and, when used for trap or polish filtration, are normally backwashed at

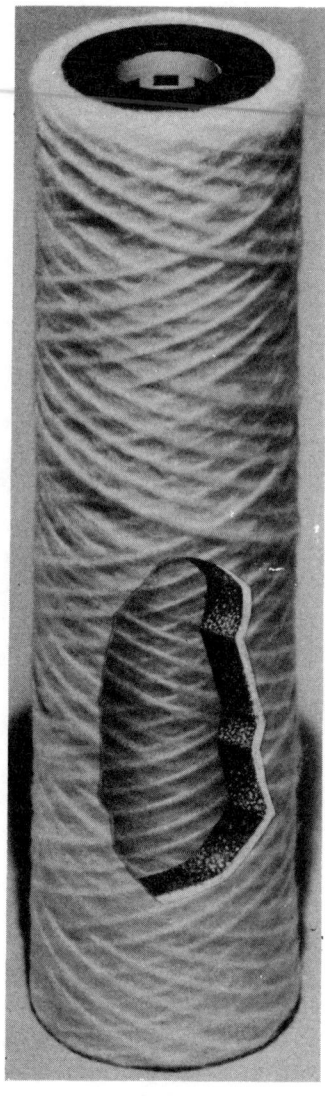

Figure 5.36 Typical activated carbon cartridge. Granular carbon is retained between two layers of wound cotton or polypropylene fiber. (Courtesy Filterite Division, Brunswick Corporation.)

regular intervals with filtrate or water to remove as many as possible of the accumulated solids. However, if the contaminants are at all slimy, gelatinous or compressible, the pores in the cartridge soon begin blinding, and only a decreasing portion of the total filter area is cleaned by each successive backwashing, so that further cleaning procedures must ultimately be used. The most commonly used procedures are ultrasonic cleaning and acid cleaning. Ultrasonic cleaning is by far the better of the two procedures, because with acid cleaning, the porous material is gradually eroded, resulting in gradually enlarged pores. The main advantages of ceramic

Chapter 5. Alfred Trumpler

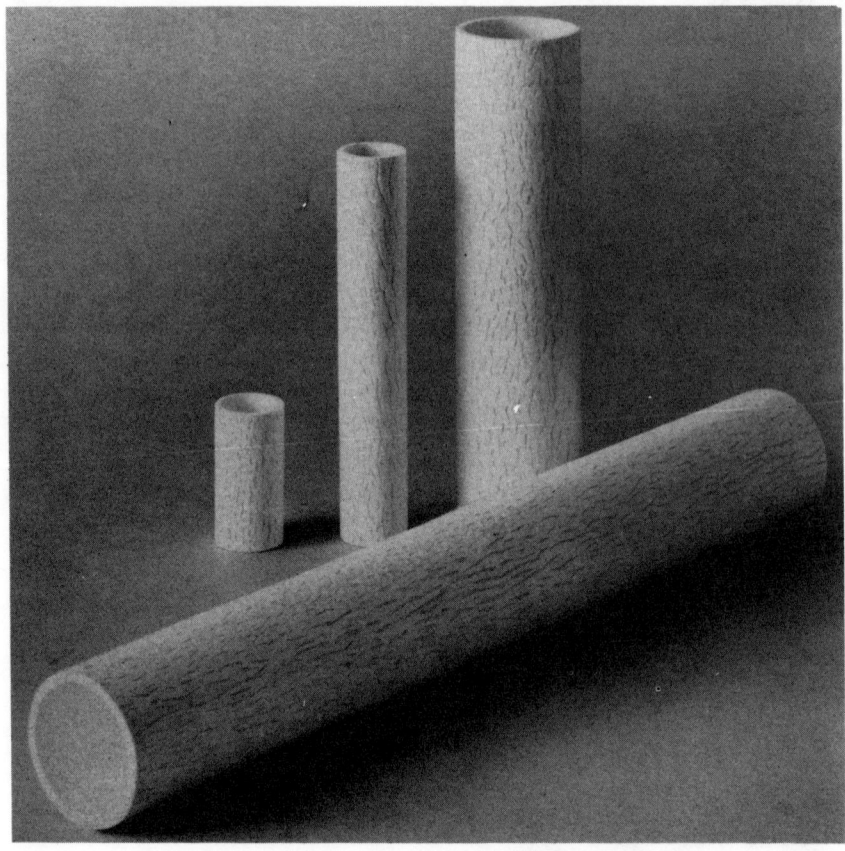

Figure 5.37 Glass microfiber filter tube, offered from 0.3—25 μm nominal retention. (Courtesy Balston, Inc.)

cartridges are their lack of migration, high temperature resistance and inertness.

Porous carbon cartridges have similar characteristics: they also can be cleaned in the same manner as porous ceramic cartridges, but are also quite brittle and easily cracked. They are used mainly in applications where inertness, high temperature or product contamination are important factors.

Porous stainless steel cartridges (Figure 5.39) are available in nominal filtration levels from 0.3—100 μm and are often used as cleanable trap filter cartridges because of their great strength and corrosion resistance. However, these cartridges also tend to blind internally after a certain period of time and must be restored by one of the two cleaning methods recommended for porous ceramic cartridges. Although stainless steel is the most widely used material for sintered metal filter cartridges, they are also available in monel, inconel, Carpenter 20, stellite, gold, silver, platinum and bronze.

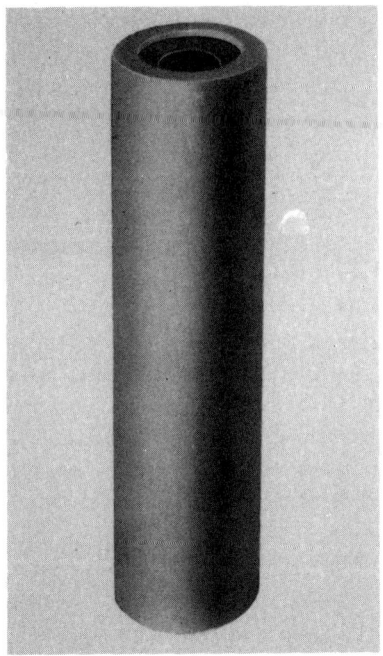

Figure 5.38 Porous ceramic cartridge shown in standard 10-in. length, but available up to 30 in. Temperature range to 900° F. (Courtesy Filterite Division, Brunswick Corporation.)

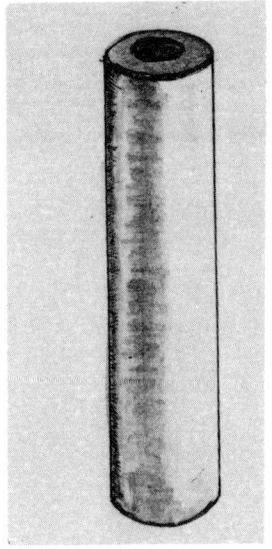

Figure 5.39 Ten-inch porous stainless steel cartridge with rubber end gaskets for sealing. (Courtesy Filterite Division, Brunswick Corporation.)

Wound wedge wire media (previously discussed under Tubular Element Filters, p. 125) in cartridge form are also available but generally are somewhat more costly than other cartridge forms. However, their inherently low blinding tendency referred to previously makes them useful in some trap filtration applications in a micrometer range of 25 and up.

Wire mesh cartridges are available both in cylindrical and pleated forms (Figure 5.40). They are produced in a variety of materials and weaves, from 20 μm up. The cylindrical cartridges can be cleaned by backwashing, but the pleated wire cloth cartridges must be cleaned by external means. One major advantage of woven wire media is its high percentage of open area, compared to many other media. One manufacturer lists in his catalog open areas from about 25–80% of total area.

Woven wire mesh cartridge applications should be carefully selected and tested, whenever possible, to determine whether the particular contaminants being removed by the mesh will separate themselves readily by backwashing, or whether a severe blinding tendency will require frequent manual cleaning. If woven wire cloth is once allowed to blind too badly, it sometimes cannot be renewed by any reasonable cleaning methods and must be discarded. In general, these cartridges should be used on low solids applications for trap filtration.

Figure 5.40 Pleated wire cloth cartridge offered in 44–250 μm range. (Courtesy Filterite Division, Brunswick Corporation.)

Various forms of woven and felted fabric cartridges are also available. These generally take the form of sleeves, which are removable from the cartridge support structure and are renewed by laundering or soaking in caustic or acid solutions (Figure 5.4). Such sleeves, usually made of nylon, polypropylene, polyester or other synthetic fabric felts, may be reused a number of times before they become blinded to the extent that they can no longer be restored to any condition approaching their original permeability, and they must then be discarded. However, since their cost is moderate, no great expense is involved.

Fabric screen sleeves are available from approximately 35 μm up, and polypropylene felt sleeves are offered in the nominal micrometer range of 5–100.

Although there may appear to be considerable savings in utilizing removable-sleeve cartridges instead of throwaway cartridges, the overall economics must be considered. Surface media, such as sleeves, will invariably become blinded and require changing many times more frequently than depth-type cartridges. Therefore, considerably more labor will be required over a given period of time for cartridge changing, and sleeve cleaning and replacement, since sleeves must be very carefully secured into the cartridge structure to be certain there are no folds or creases to permit by-passing of dirty fluid.

It cannot be said that there are great advantages either way when comparing throwaway and cleanable cartridges. Generally, the same filter housings are built to contain both types, in a broad size range from 1 gpm (3.78 l/min) to several thousand gpm. Both types are offered by many manufacturers in a wide nominal micrometer range from 0.2 up to several hundred. Both types may be equally effective in removing contaminants from a liquid.

As has been stressed throughout this chapter, proper application of filters must be considered from many standpoints, and whenever laboratory or pilot-scale testing is possible, this should be done, because it is difficult to predict behavior of any cartridge, either throwaway or cleanable, with a given type of contaminant in a given liquid. Quite often, a backwashable cartridge will prove to be much more economical in an application that tends to give very short life with wound or molded cartridges. On the other hand, the contaminant removed by throwaway cartridges is discarded as a dry waste along with the cartridges, whereas backwashing of permanent cartridges results in a slurry that must be disposed of in sewers or to a waste treatment system. Furthermore, the frequent handling and cleaning of many types of cleanable cartridges requires a certain amount of costly labor and sometimes results in cartridge damage or breakage.

Often, the relative differences in migration and leachable impurities are the major considerations that dwarf direct cost and other parameters. These two characteristics of various filter media will be discussed in a later section.

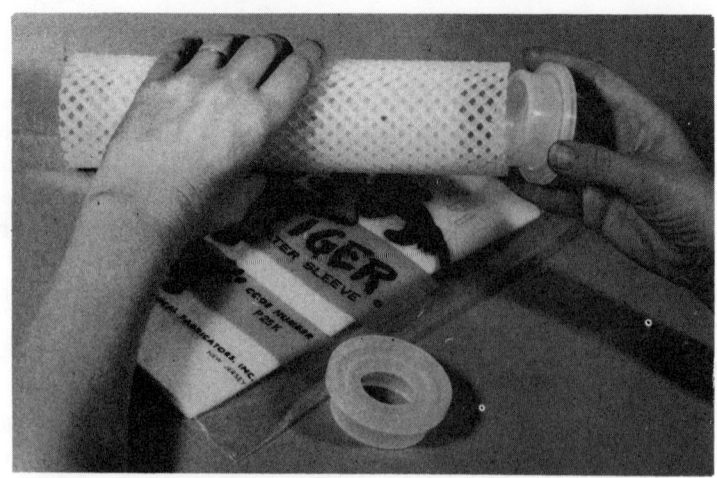
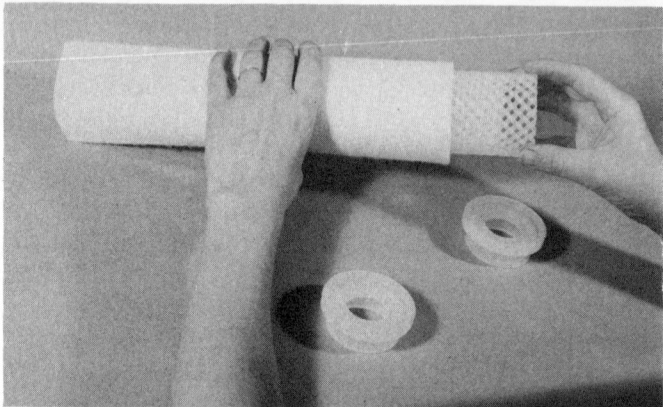
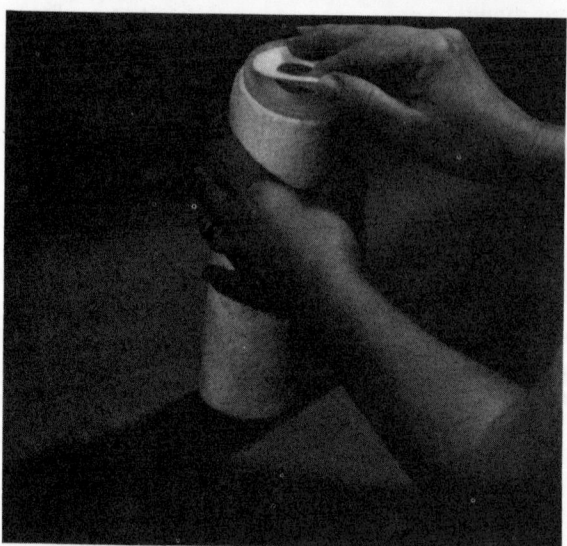

Figure 5.41 Replaceable polypropylene felt sleeve cartridge showing steps in sleeve replacement. (Courtesy Technical Fabricators, Inc.

Process Equipment Series Volume 1

TUBULAR ELEMENT FILTERS

Single Housing — Inline

Many manufacturers offer a line of inline pressure filters (Figure 5.42) which contain one tubular filter element per housing and which offer any of the media configurations described in the previous section.

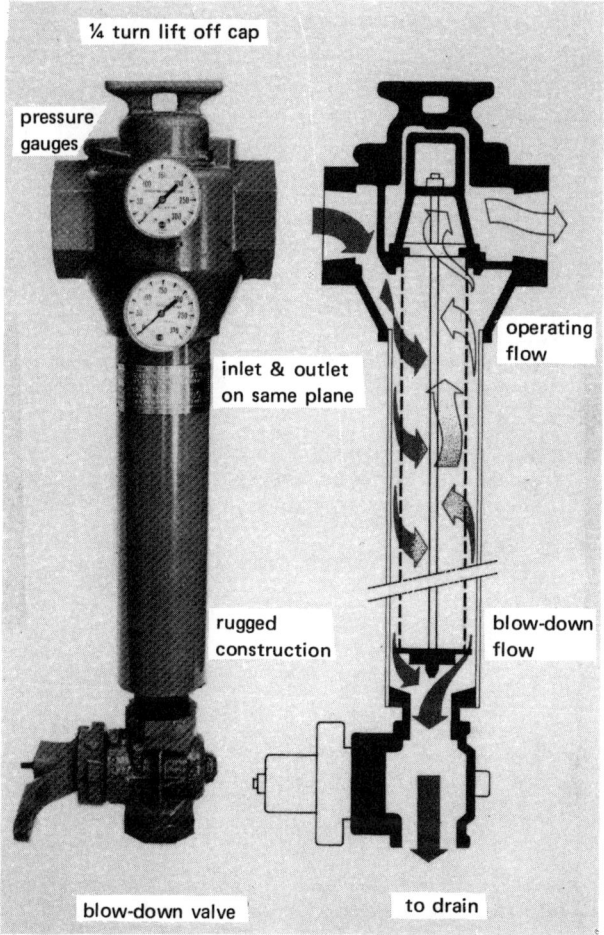

Figure 5.42 Typical inline filter utilizing filter element that can be removed or backwashed in place. (Courtesy Albany Engineered Systems.)

Although there are, of course, slight variations in design by each manufacturer, they all consist of a cylindrical housing designed to withstand pressures of 150, 300 and 1,000 psi (10.5, 21, 70.3 kg/cm^2) or more, and these housings are designed internally to take an element of some sort, from 2½ in. (6.4 cm) diameter by 10 in (25.4 cm) long to 3 in. (7.6 cm) or 4 in. (10.2 cm) diameter by approximately 40 in. (102 cm) long. Flow ratings go up to approximately 300 gpm (1,100 l/min) with retention down to 1 μm.

These elements can be removed manually and cleaned by washing off, scrubbing, soaking or whatever method works best (some manufacturers provide quick-opening top closures for rapid element removal). They can also be cleaned at regular intervals by backwashing in place, and one manufacturer offers an optional compressed air dome which provides a shock backwash claimed to aid in removal of "tenacious solids," which create a strong blinding tendency (Figure 5.43).

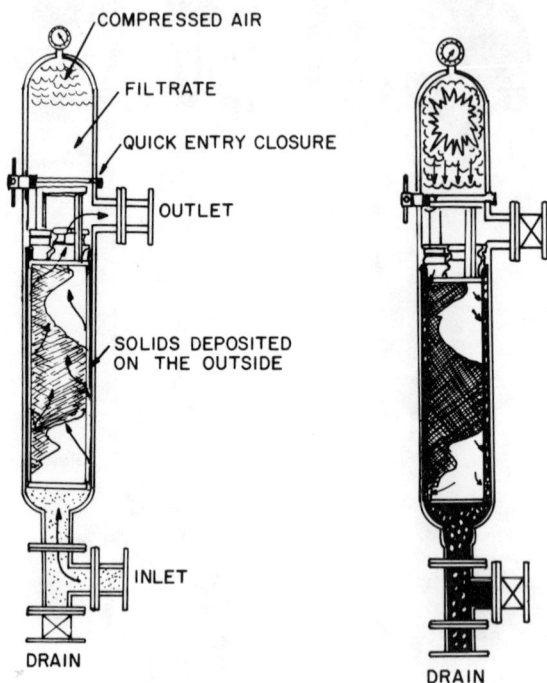

Figure 5.43 Inline filter with air dome for shock backwash. (Courtesy Filtration Systems Division, Duriron Corporation.)

Continuous—Flow Dual and Multiple Filters

When a filter must be in service continually, and backwash cleaning is suitable, a dual or multiple-housing system (Figure 5.44) is used so that one element at a time can be cleaned either by manual removal or by backwashing, allowing the other

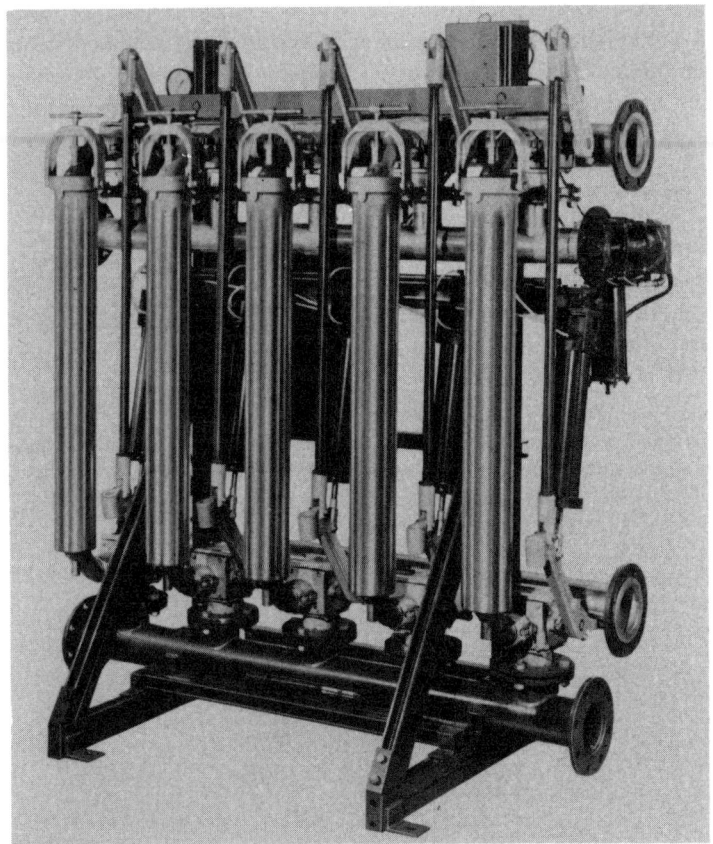

Figure 5.44 Automatic multiple backwash filter. This unit provides continuous automatic operation. (Courtesy Ronningen-Petter Co.)

housing or housings to handle the line flow without interruption. Flow capacities of such multiple housings are almost limitless because several housings can be manifolded into one bank, and any number of banks can be manifolded together in parallel.

Although it is possible to operate such dual or multiple systems manually, where operators are constantly on hand to observe pressure differentials on the gauges and open backwash valves as necessary, these systems lend themselves ideally to automation. A relatively inexpensive automatic control package provides either pressure differential or timer initiation of the backwash cycle, which consists of backwashing each element for a predetermined period of time (usually 5–60 seconds) with an external source of water or with filtrate from the other filters, then automatically returning to normal filtration.

However, with any continuous system, regular observation of pressure differentials is essential to prevent the possibility of all elements blinding simultaneously. When this occurs, even after the backwash cycle is completed, the resistance to flow of blinded filter media is too great to provide the required system capacity.

Even in an automatic system, therefore, pressure differentials and filter media condition must be checked regularly.

BAG FILTERS

Although some types of bag filters (Figure 5.45) have been on the market for many years, the bags have been used mainly as septa, with filteraids actually accomplishing the desired filtration. In recent years, a new type of bag filter in which the bags are the filter media, has found a place in the market.

These filters consist of three parts:

1. a simple pressure vessel designed to hold one or more bags;
2. a perforated or wire mesh retainer basket, which seals tightly in the filter vessel; and
3. one or more disposable round filter bags, approximately 7½ in. (19.1 cm) diameter by 18–31 in. (45.7–78.7 cm) in length.

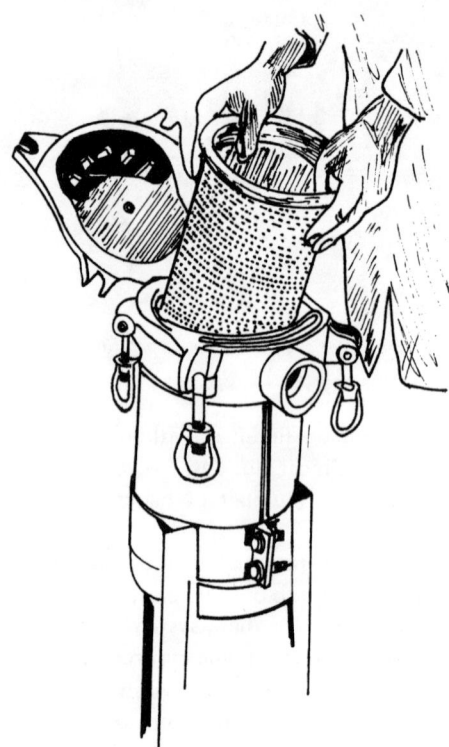

Figure 5.45 Bag filter vessel (bag not shown). (Courtesy FSI, Inc.)

Bags are generally available in a nominal micrometer range of 5–200, although one manufacturer offers a range of 1–800. Materials offered are viscose rayon, polyester, nylon, polypropylene and wool/silk felt, and nylon, polypropylene, polyester and saran monofilament woven mesh fabrics. Optional bag covers are available to provide strength and reduce fiber migration.

Bags generally provide between 3 and 5 ft^2 (0.28 and 0.46 m^2) of filter area each, and a single bag may be rated at flows up to 40 gpm (151 l/min). Most bags are basically single-layer surface filters and tend to blind fairly quickly if used for slimy, compressible solids, but multilayer bags, which provide some degree of depth filtration, are also available. The bag filter is relatively simple and fool-proof, and bags are inexpensive and can be changed quickly.

Once again, no generalization can be made concerning bags versus throwaway or cleanable cartridges. Where either can be used in terms of micrometer filtration requirements, material compatibility, etc., the decision as to which type filter will provide the lowest cost per gallon must be made on the basis of a small test filter of each type installed and evaluated on a slip-stream basis, or the process engineer's past experience and judgment concerning the amount and nature of the contaminants.

In the absence of testing, the only guide for application of bags versus cartridges is that crystalline, noncompressible solids (for example, fine sand) will provide relatively low resistance to flow; that is, they exhibit low blinding tendencies, and with the large amount of area and solids-holding capacity of the felt bags, considerable amounts can be retained in bags before disposal is necessary, and cost per gallon filtered may be very low.

At the other extreme, compressible, slimy contaminants like yeast, algae, many metal hydroxides and undissolved gels, will tend to blind a bag surface almost completely with an infinitesimally thin layer. Bag life may be very poor and cost per gallon filtered very high compared to depth cartridges.

PAD OR SHEET FILTERS

Although seldom used in the chemical industries, pad or sheet filters have wide applications in pharmaceutical and alcoholic beverage industries for trap and final polish filtration, often just before sterile, membrane-type filters to obtain maximum life of costly absolute membranes.

The filters used for pad or sheet filtration are very much like the plate-and-frame filters described previously; however, the majority of them have no frames between the plates because there is rarely a need for cake space in this type of filtration. Also, a number of pad filter designs offered today are externally ported.

Because of the type of service in which they are used, these filters are usually made of stainless steel, although sometimes plastic plates are used to reduce initial cost. They are made with highly machined surfaces and sanitary construction throughout. The ability to thoroughly clean and sterilize them is very important.

The filter media (pads) (Figure 5.46) are relatively thick (3–4 mm) and quite stiff. They have for many years been made primarily of a blend of cellulose and asbestos fibers, but due to FDA regulations and the general concern about handling asbestos and glass fibers, most manufacturers are offering nonasbestos pads made of a combination of cellulose fibers and nonfibrous materials, such as certain siliceous materials and wet strength resins.

Figure 5.46 Various shapes and sizes of filter pads used in pad and sheet filters. (Courtesy AMF Cuno Division.)

They are machine-formed to provide various nominal permeabilities in the range of approximately 0.5–30 μm. They are depth filters and retain many colloidal particles, yeast and bacteria, which must be removed from pharmaceutical liquids and beverages.

Pad filters have, to some extent, the same disadvantages mentioned in earlier sections under plate-and-frame filter presses: product leakage and high cleaning and set-up labor.

Pad filters are now also available as enclosed pressure vessels with the cellulosic pads formed into accordion-type cartridges (Figure 5.47). Although the pads are more costly in this form, this type of pad filter has successfully eliminated the two major disadvantages of the plate-type pad filter.

EDGE FILTERS

Edge filters (Figure 5.48) contain all-metal filter cartridges, which are made of wheel-shaped discs and spacers stacked on a rotatable shaft. Stationary cleaning blades are mounted next to the cartridge stack and project into the spaces between

Figure 5.47 Accordion-type pad filter cartridge. (Courtesy AMF Cuno Division.)

discs, so that when the cartridge is rotated, the cleaning blades clean out the filter slots. The solids accumulate in the filter sump and can readily be blown out through a bottom drain. The cartridge can be rotated for cleaning either manually by means of an external handle or by motor.

These units are more often used as strainers than as filters, but slot openings as low as 0.0015 in (0.038 mm) (36 μm) can be obtained. Capacities range from several gpm up to 100 gpm (378 l/min) or more. Their instant cleanability feature gives them great advantages over screens and basket strainers and they are ideal for use as prefilters for removal of a large percentage of contaminants before a finer cartridge or other type of polishing filter.

Edge filters cost more than most other filters; they are not usable for fine filtration; and if they are misapplied in situations involving heavy, slimy solids, and

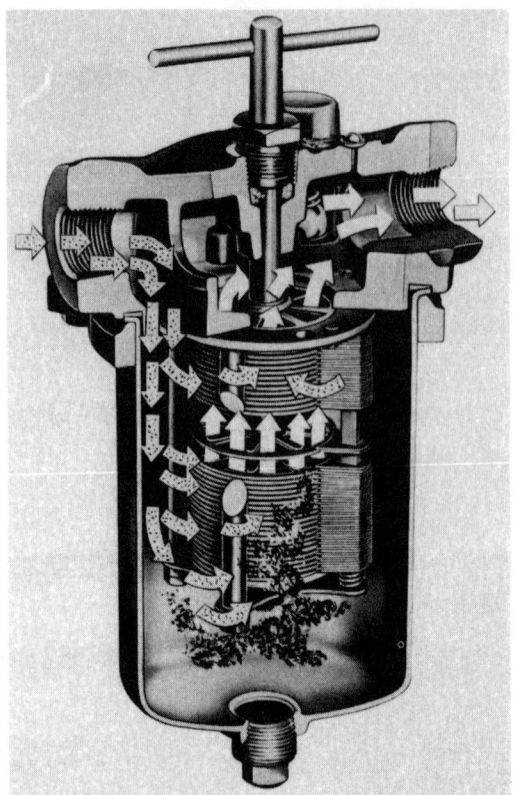

Figure 5.48 Edge filter cutaway showing flow pattern and cleaning method. (Courtesy AMF Cuno Division.)

not cleaned frequently enough, it becomes very difficult, if not impossible, to rotate the element without breaking the shaft. When this occurs, the elements must be disassembled manually and cleaned.

DEEP BED FILTERS

Deep bed filters (Figure 5.49) as their name implies, utilize a bed of filter media from 1–10 ft (0.3–3.0 m) thick, enclosed within a pressure vessel, and incorporate means of backwashing the filter media as necessary.

Single and multiple filter systems of this type are generally high capacity, in the range of 20–50,000 gpm (75.60–189,000.00 l/min) and are used for such applications as tertiary sewage plant effluent, industrial intake water supplies, final plant effluent treatment and large-scale clarification of process waters.

They are equipped to handle solids loads of up to 1,000 ppm and, in many cases, effluent suspended solids can be held to 1 ppm or even less. In most cases, inlet

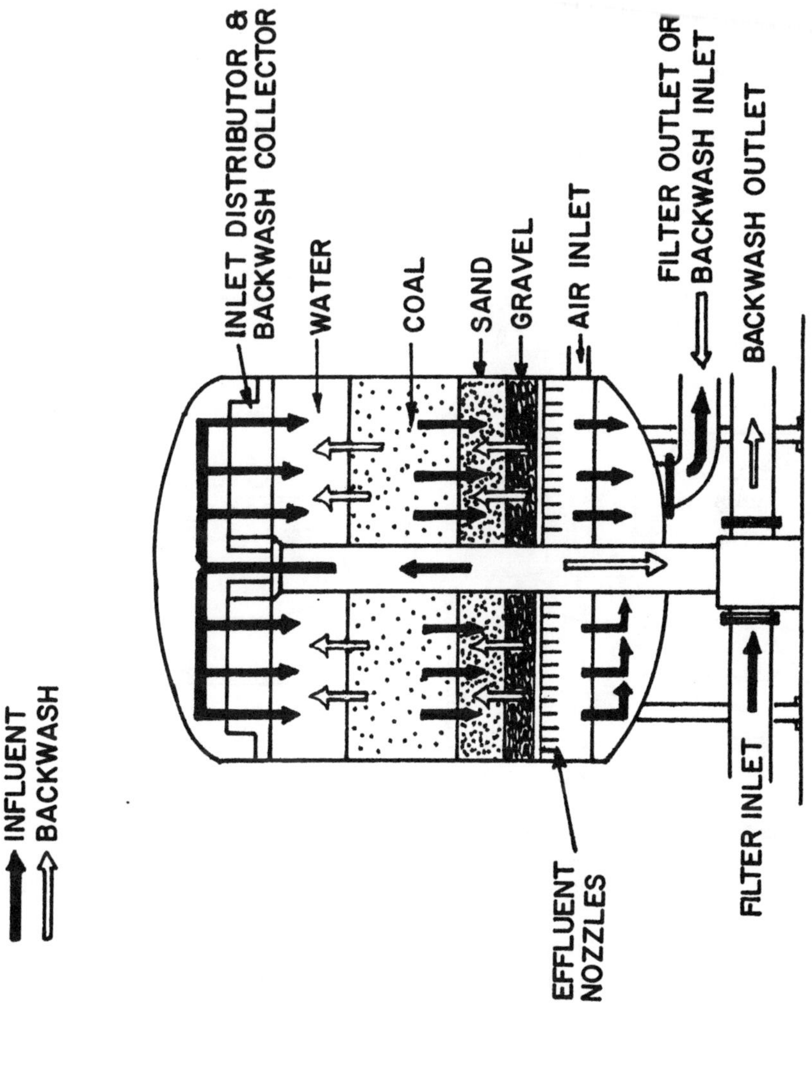

Figure 5.49 Diagrammatic cross-section of one type of deep bed filter utilizing coal, sand and gravel filter media. (Courtesy DeLaval Turbine Inc., Condenser & Filter Div., Florence, NJ.)

solids loads are far less than 1,000 ppm, and effluent requirements are more normally in the range of 5—15 ppm.

Filter beds are made up of a variety of media, including gravel, graded sand, ferrosand, anthracite, activated carbon (Figure 5.50) and polymer resin media. The large vertical pressure vessels contain distributors for inlet fluids, as well as distribution arrangements for the backwash water. The key to success in this type of filter is the effectiveness of dirt and oil removal from the filter bed with a minimum of backwash water. A number of manufacturers of this type of filter claim backwash water volume of 2% of filtered water. At least one manufacturer utilizes air and water together for effective cleaning of the filter media.

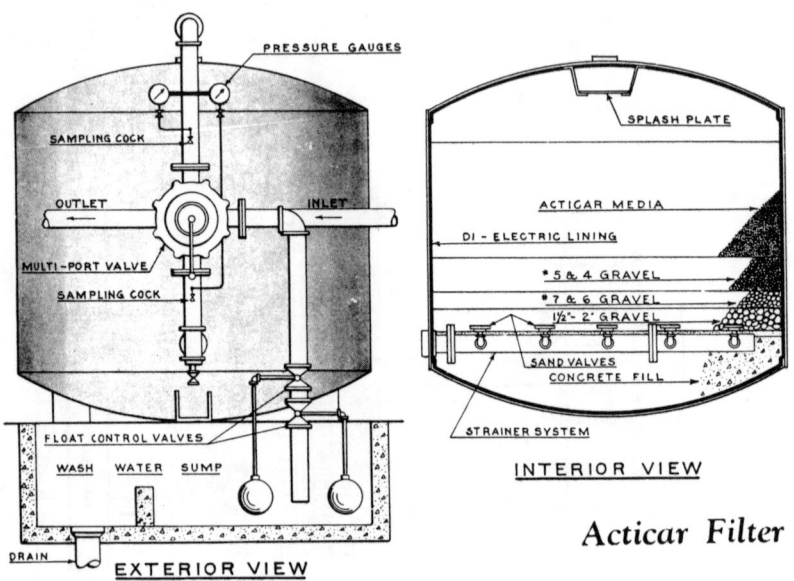

Figure 5.50 Interior and exterior view of an activated carbon bed filter, used generally for removing organic impurities, chlorine, sulfur, algae and other taste- and odor-causing materials from water. (Courtesy Hungerford & Terry, Inc.)

Forward flows of up to 20 gpm/ft^2 (817 l/min/m^2) of bed area are possible, and backwash flows are often greater than this. Maximum pressure drops across these deep beds are in the range of 10—25 psi (0.7—1.8 kg/cm^2).

Influents are often pretreated with polyelectrolytes or other coagulants to promote agglomeration of fine solids into a more filterable floc. These filter systems are often fully automated in view of the relatively short filtration cycles.

When these filters are used as carbon bed filters, utilizing activated carbon, they become more than solids removal devices, since they are usually used to adsorb chlorine, sulfur, phenols, and other dissolved products, which cause bad tastes or odors in water.

Deep bed filters are limited in their application by their size, materials of construction and by the fact that the amount of backwash water may present a problem in many plants, from a disposal standpoint. Also, most media used in this type of filter do not have as low a micrometer particle removal capability as precoat filters. However, in spite of these limitations, they fill a vital need in a number of relatively large-scale industrial applications.

ABSOLUTE ("MEMBRANE") FILTERS

Although the chemical industry itself does not have frequent need for microfiltration, with absolute particle removal cutoff, there are uses for this type of filtration in associated process industries like pharmaceutical and beverage, so a very brief discussion of absolute filters and filtration is appropriate in this chapter.

Many references have been made so far to "nominal" micrometer ratings. This is because most of the filter media discussed so far have been depth filters to a certain extent, and by the very nature of filtration through depth media, they will remove many particles of a wide range of sizes; however, there is no one point at which it can be said that every particle above that micrometer size will be retained.

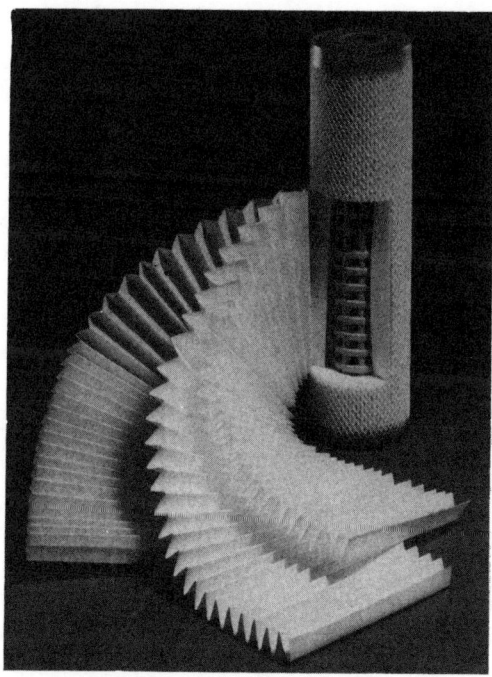

Figure 5.51 Pleated membrane absolute filter cartridge. (Courtesy Gelman Instrument Co.)

Chapter 5. Alfred Trumpler

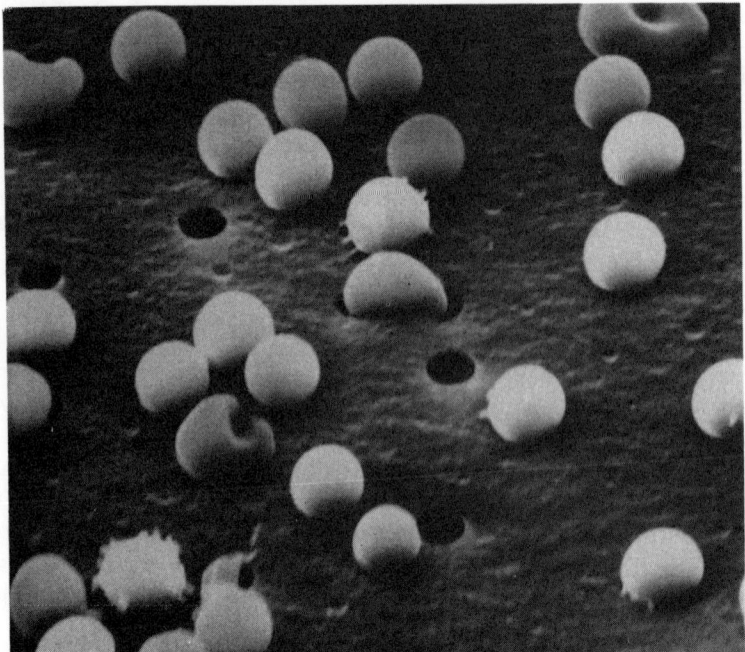

Figure 5.52 Photomicrograph showing red blood cells being retained on an absolute membrane filter medium. (Courtesy Nuclepore Corporation.)

Absolute filter media (Figure 5.51) are made to perform as screens with very accurately controlled opening or pore sizes, and anything larger than the maximum pore size must be retained on the surface of the media.

These media are made of a number of materials; cellulose, esters, polyethylene, cellulose acetate and triacetate, polyvinylidene, fluoride, cellulose (rayon), cellulose nitrate, Teflon, polycarbonate and others. Most of these are referred to commonly as membranes. Thickness of most absolute filter membranes (Figure 5.52) ranges from 5–125 μm and absolute ratings are from 0.15–12 μm.

They are normally used as flat discs in special highly machined and readily sterilizable filter holders (Figure 5.53) but an increasing number of manufacturers are now offering membranes in cartridge form, both cylindrical and pleated.

Since both hardware and media are quite costly in comparison with the pressure filter types discussed so far, it is generally necessary to do an excellent prefiltration on any liquid requiring microfiltration to utilize minimum size hardware, but mainly to get as long a membrane life as possible. The best way to do this is by combining the good properties of both absolute surface filters and nominal depth filters, which can be used to retain the great majority of the contaminants.

The ultimate choice of prefilters requires careful consideration of filtration efficiency, dirt-holding capacity and cost. To further reduce the total dirt load on the filters selected, it is often desirable to consider additional pretreatment, such as

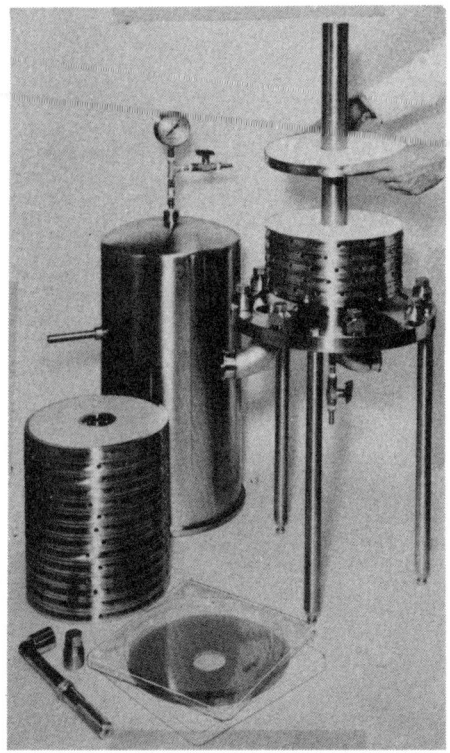

Figure 5.53 Removing a filter disc and support plate from a multiple-plate, membrane filter housing. (Courtesy Millipore Corporation.)

settling, agglomeration, flocculation, carbon treatment and such nonfiltration processes as ion exchange and reverse osmosis.

MISCELLANEOUS PRESSURE FILTER TYPES

Mud Discharge Filters (Figure 5.54)

At least two pressure filter designs exist that produce a "mud" discharge of the accumulated solids in the filter. One of these is a vertical cone bottom design, which will concentrate (dewater) sludges from 500 ppm or more down to about 20% dry solids by weight. Available from about 100–500 ft^2 (10–50 m^2) of filter area, it is normally designed to operate automatically.

Using vertically positioned, fabric-covered filter leaves or tubes, this filter relies on the specific gravity of the filtered solids being sufficiently greater than the liquid to allow them to settle readily. When the filter leaves or tubes become loaded with solids, as determined by pressure drop or a cake thickness detector, an external

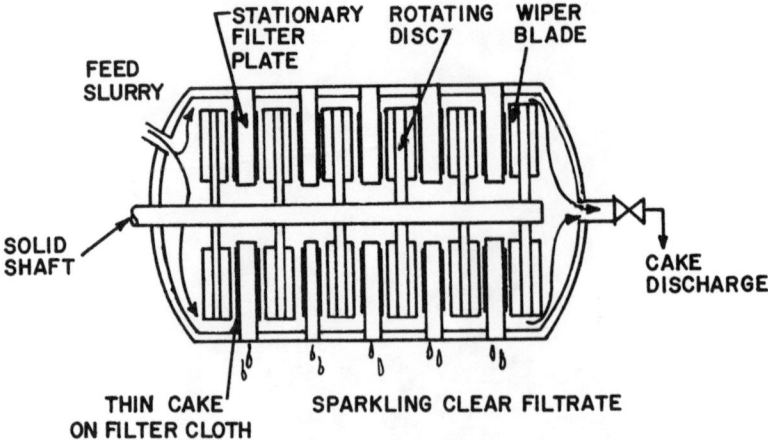

Figure 5.54 Simplified schematic of one continuous mud discharge filter. Slurry increases in concentration as it moves from inlet to outlet. (Courtesy Artisan Industries, Inc.)

vibrator is actuated and the filter cake is dislodged in a matter of seconds. While filter feed is stopped, the dislodged cake is allowed to settle in the cone bottom. When this settled cake reaches a desired consistency, as determined by a density detector, the mud discharge valve opens and the mud is extruded from the filter. Filtrate quality is said to be in the range of 1–2 ppm of solids.

The other design is a closed, horizontal pressure filter, which is made up of a series of vertical, cloth-covered round plates and which is available in sizes from 3–200 ft^2 (0.28–18.6 m^2).

Its operation is continuous, not cyclical like most filters. Its manufacturer claims a unique principle of operation: filtration through a thin, permanent cake of the contaminants only (no filteraids).

During operation, a thin cake of the contaminant is permitted to form on each filter plate and is kept at constant thickness. The cake becomes the filter medium over which the feed material washes continuously. The progressively thickening feed slurry moves through the various filter chambers with some clean liquid being removed at each successive stage. By the time it reaches the discharge end of the filter, the cake is said to be of a toothpaste-like consistency and is discharged.

In both of these filter designs, testing by the manufacturer to determine characteristics and behavior of the slurry is recommended, since a number of variables can greatly affect their performance and operation.

CONCLUSIONS

A Few Other Selection Parameters

Several characteristics of filter media and septa, which have not been discussed so far may actually be the overriding considerations in selecting filters for certain services. Whenever one of these characteristics becomes a severe limitation, all other parameters become secondary.

Migration and Bleedthrough

The terms migration and bleedthrough in filtration language refer to the very undesirable characteristics (possessed by just about all filter media and septa) of allowing some of themselves to remain in the filtrate.

Migration has particular reference to solid media like cartridges, paper, porous stainless steel, etc., whereas bleedthrough refers mainly to powder or fibrous filter media, such as filteraids.

Various solid media have varying tendencies to migrate downstream and, in some applications, this migration tendency may be of such importance that some types of filter media may have to be discarded from consideration.

For example, most throwaway cartridges, whether wound, molded or pleated paper, tend to migrate a certain number of fibers, particularly in the first few gallons after they have been put into service. To reduce the number of loose bits of fiber that tend to migrate, core covers and other built-in fiber-free layers of various materials are used, but in most cases they are not 100% effective. In certain types of applications, like photographic film and paper coatings, migration of even a few fibers can have a rather serious effect. The user should be aware of these potential problems in making his selection.

Bleedthrough of filteraids and fine carbon (when this is present) is a potential problem with most precoat filters. Particularly when woven wire cloth is used as the filter septum, there is a tendency for a certain amount of fine diatomite or perlite to remain in the filtrate, and precautionary measures must be taken to overcome this tendency, particularly in process liquids where even a few ppm of filteraid could be a problem.

Initial selection of filter septa has a bearing on bleedthrough tendencies, but often the problem is resolved by utilizing fibrous cellulose filteraids as a precoat layer on the septum being used, or by adding an inexpensive trap filter downstream to pick up any filteraid bleedthrough, or by both.

Media Compatibility and Leachable Impurities

It goes without saying that one of the primary requisites to be used in filter media selection should be compatibility with the liquid being filtered, not only from temperature and corrosion standpoints, but also from the viewpoint of adding contaminants to the filtrate. A few examples follow.

Liquids that tend to react with or hydrolize cellulose cannot be handled in a filter utilizing paper in any form. Chemicals sensitive to iron contamination cannot be handled even with stainless steel equipment or media. Liquids sensitive to silica or any of the other minor impurities in diatomite or perlite cannot be handled with filters or media that require filteraids. Many wound cartridges have a certain amount of an antistatic finishing agent in the fiber roving, and these finishing agents leach out in both aqueous and solvent-based liquids to a slight extent. The melamine and phenolic resin impregnants in many types of paper and fiberglass media are incompatible with some organic and inorganic chemicals.

It is essential, therefore, for the user to consider all these limitations, as well as the parameters laid out at the beginning of this chapter for equipment and media selection.

SUMMARY

It is by now apparent to the chemical plant designer and engineer that the selection of liquid pressure filters is a relatively complicated process. With all the equipment types and septa and media configurations we have discussed, it would seem that taking a close look at Figure 5.2 and using the process of elimination is the best approach.

When the selection has been narrowed down to several types, it is best to make the final decision based on small-scale pilot testing (many of the larger filter manufacturers have pilot filter rental programs for this purpose) and economic comparisons of both capital and operating costs.

CHAPTER 6

PARTICLE CLASSIFICATION USING FELT STRAINER BAGS

ARTHUR C. WROTNOWSKI*
GAF Corporation
Greenwich, CT

INTRODUCTION AND THEORY

Felt strainer bags are especially well adapted for removing microscopic-sized particles, *i.e.*, 1–50 μm (see Table 6.1 for a definition) from solutions, slurries, dispersions, emulsions, hydrocarbons, fatty acids, solvents, waste streams, exotic liquids, latexes, etc., *ad inifinitum*.

Table 6.1. Definitions

Nominal Rating — 90% removal of specified particle size, μm
Straining — Removal of particles larger than the medium pore size (1)
Filtration — Removal of particles smaller than the medium pore size (1) thereby employing the filtration mechanisms:
       1. Streamline flow
       2. Diffusion
       3. Gravitation
       4. Inertial
       5. Electrostatic
Snap Ring Bag — a filter bag having a built-in flexible ring at its top to provide easy installation in the Open System and the closed pressure vessel system (Figure 6.3)
Microscopic Size Particles: 1–50 μm
    For the purpose of description in this section the particles to be classifiled by felt strainer media canbe referred to as microscopic-sized particles. This is based on the physical limit of an optical microscope's resolution, about 1 μm, and the smallest object visible to the naked eye, approximately 50 μm.

The difficulty in handling microscopic-sized particles is the physical problem of catching and collecting them on a practical basis. Screens, woven wire cloth or other flat, two-dimensional structures are not suitable for handling these small particles. Harris (2) pointed out some limitations of screen sizing. For example, the "difficult size," where blinding of the screen and the rate of classification is critical, occurs when particles are the same size as the screen pore size. Furthermore, Harris sets 500 μm and larger as the trouble-free size for screen sizing.

---

*Present address FSI Filter Specialists Inc., Michigan City, Indiana.

On the other hand, felt strainer media, pore size 1–200 μm and made from a variety of textile filters employing a low-density, high--voidage design, are found to be quite suitable for classifying small particles. Felts are manufactured by a physical interlocking of individual fibers to form a uniform, well-constructed material. The felting process creates fiber orientation in three directions to make a "homogeneous" fibrous structure, approaching a true isotropic material, *i.e.*, physically uniform in three dimensions. Lauterbach (3) introduced the process of needling synthetic fibers to form felt.

A felted strainer material is a homogeneous, uniform, interlocked fibrous structure usually composed of one single fiber diameter and one single polymeric fiber. Felts of this type are statistically responsive to adjustments in construction, *i.e.*, given the fiber diameter and felt density the effective felt strainer pore size can be predicted with good accuracy by means of a mathematical model (4, 5).

Crimped fibers are used to form lofty structures, which have a large pore fraction, approximately 85%, an index of solids-holding capacity. The individual fibers and the pores as formed are relatively strong and are uniformly spaced throughout the felt. A given felt construction tends to entrap one given size particle. The smaller particles pass by the entrapped particle and the larger particles are retained by gross straining action, somewhat as shown in Figure 6.1 (5). The schematic micrograph of Figure No. 1 is based on microscopic examination, microphotographs of collected particles and by visual examination of 100-μm particles retained in 100-μm strainer felts.

"Fines" pass by collected particles, resulting in large solids classification capacity

When classifying, particle leakage of controlled size is desirable

Figure 6.1 Schematic of entrapment and classification mechanism.

The felt strainer process, which was commercially started in 1964, (4) was well established by 1978. End uses and application continue to expand, thereby identifying felt straining as an engineering tool available to the art and science of liquid-solids separation. The relative position of various solids separation processes are given in Figure 6.2.

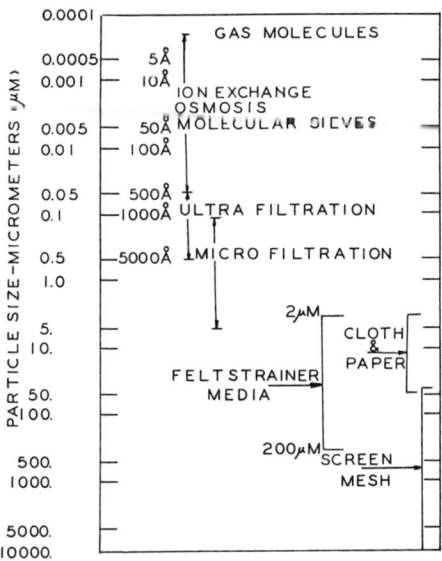

Figure 6.2 Relative positions of various solids separation processes.

PRACTICAL DESIGN APPROACH

A felt bag has been found completely satisfactory as a means to use the particle classification capacity of felt strainer media. The felt material is made structurally strong by means of an internal synthetic fiber scrim, with apertures large enough (1,000 μm) so as not to interfere with the straining action. The bag is fabricated with a rounded end to eliminate stress concentration and to accommodate hydrostatic pressure requirements. The straightforward and familiar bag configuration is readily acceptable by users.

The felt bag has been made interchangeable between an "open" and a "closed" straining system, and the bag is used as a module to develop different-sized equipment. Notice in Figure 6.3 a perspective view of the equipment line. Comments on each style of strainer follow:

In practice, the felt cone gives a good estimate of the level of straining accomplished by the range of felts available. As a rule the cone, when used as a bench top test, indicates more leakage than would be observed in production. This is because the felt cone test is performed on clean felt, *i.e.*, no solids or internal cake is yet present. Figure 6.4 demonstrates how the cone can be used to estimate production performance.

The plain bag shown in Figure 6.5 demonstrates a "rough and ready" method to

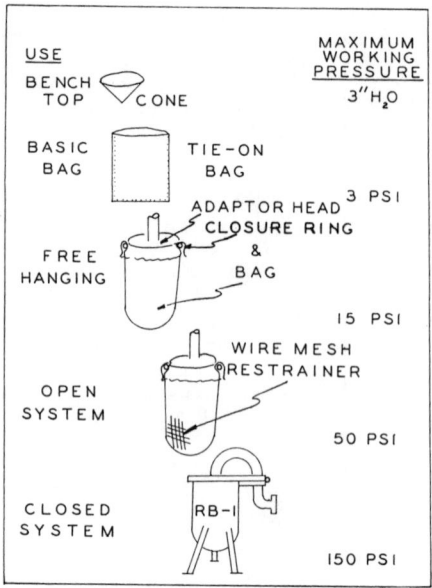

Figure 6.3 Strainer bag concept.

Figure 6.4 Felt filter cone.

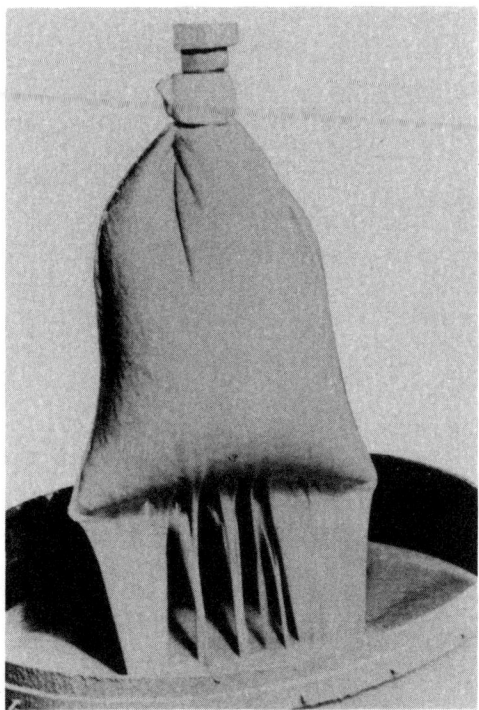

Figure 6.5 Basic (tie-on) bag.

use the felt. One can see that the bag material is strong, which is due to the previously mentioned internal synthetic fiber scrim. In spite of the elementary equipment, as shown, quality straining of particles in the microscopic range are quite readily achieved.

The Snap-Ring®* bag with rounded bottom and adaptor head has proven to be efficient, reliable and easy to install and replace. The bag can be operated at 10–15 psi as an open system (Figures 6.6 and 6.7). The wide range of micrometer sizes and the various fibers employed make a versatile open system with Snap Ring bags readily available as an economical, quality straining method requiring a minimum of equipment. The process is immediately suitable for pilot plant investigation, moderate-size production, a roughing filter, or to provide quality control of liquid streams.

By placing a wire mesh restrainer over the Snap Ring bag (Figure 6.8) the open system applications are greatly expanded by increasing the operating pressure to 50 psi.

---

*Snap Ring is a registered trademark of the GAF Corporation.

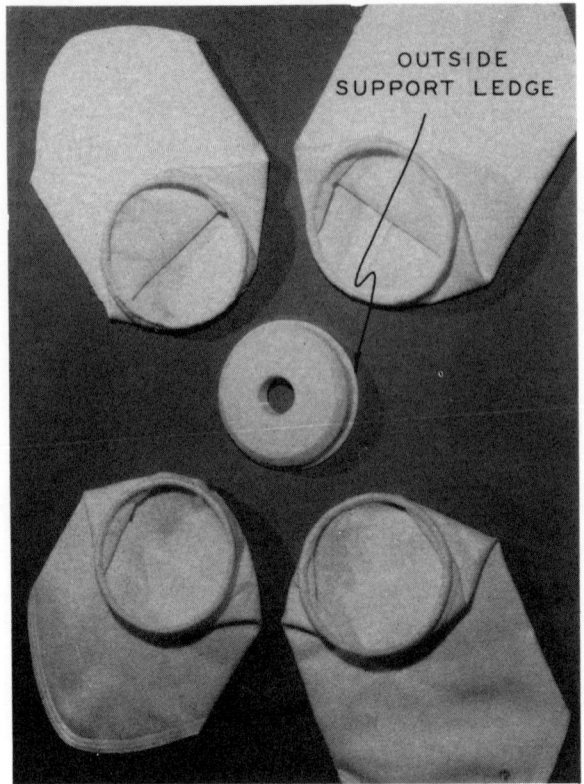

Figure 6.6 Adaptor head and Snap-Ring bags.

The RB hardware (Figure 9–20) can fit into practically all chemical process industries where some particle size removal is needed. The chemical stability of the synthetic fibers used and the precise straining available in the microscopic range of particles makes the units most useful to the chemical industry.

OPEN SYSTEM DETAIL AND OPERATION

There are two limitations to the open system of felt straining:
1. *Volatile liquids* are not satisfactory because of the worker's safety and because of product loss.
2. *Continuity of Piping.* The strained product does not simultaneously move along to the next operation.

However, where these conditions are allowable or if an alternate arrangement is suitable, there are economic and versatility advantages to the open system as follows:

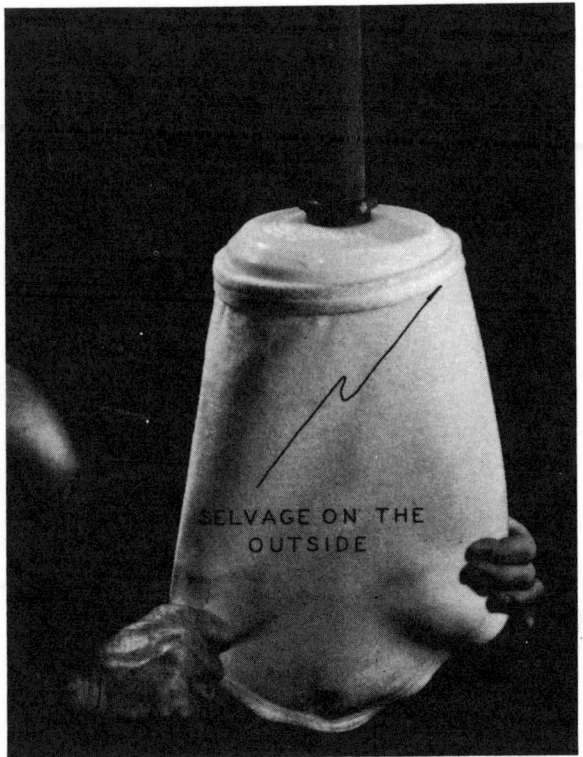

Figure 6.7 Snap-Ring bag installed.

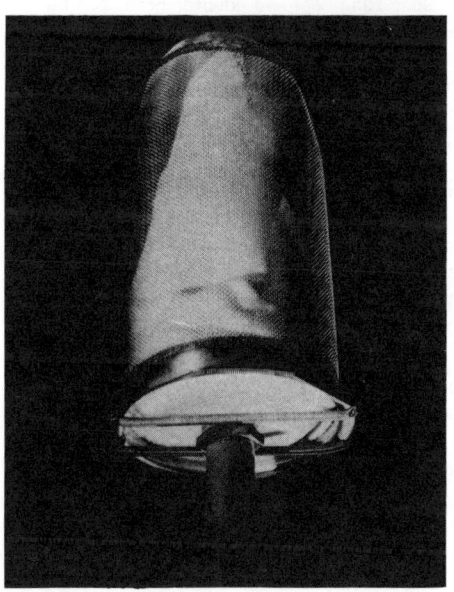

Figure 6.8 Snap-Ring bag restrainer.

Chapter 6. Arthur C. Wrotnowski

Figure 6.9 Model RB-X, 1 ft² of filter felt area.

1. *Gravity Feed.* The strainer bag media as a class are characterized by high permeability, which is synonymous with low pressure drop when in use. A 10-ft liquid head, approximately 5 psi, will generate a clean felt theoretical flowrate of 100 gpm/ft² of average strainer medium (Figure 6.21).
2. *Manifold Piping.* An indefinite number of Snap-Ring bags and adaptor heads can be mounted in parallel by means of pipe tees, ells, etc., expanding on the single bag shown in Figure 6.7. An open trough collection system would be functional.
3. *Complete Chemical Stability.* Based on that information in Tables 6.2 and 6.3, almost all chemicals can be handled with complete chemical stability.
4. *Reusable Bags.* In most cases the felt bag does not "blind," *i.e.* become irrelieveably clogged. However, it is usually more economical due to labor cost to discard the used felt bag and collected solids rather than to clean it up.
5. *High Pressure.* Using a stainless steel wire restrainer will increase the cycle life usually 3 to 4 times by increasing the operating pressure drop to 50 psi. Standard sizes No. 1 (3 ft² and 2 (6 ft²) are available. The strainer basket of Figure 6.8 should be installed a bit "high" so that as the bag fills out the restrainer basket orients downward to effectively restrain the taut strainer bag.

From an operational point of view all felt strainer media have a very low pressure drop when new and clean (see Table 6.2 for actual air permeability values and Figure 6.21 for a conversion graph giving liquid flowrate. For example:

Table 6.2 Rayon Viscose
25 μ m rating   250 cfm/ft²/½ in H_2O

Figure 21

250 cfm ft2/½ in H_2O at 800 Centipoise   4.0 gpm/ft^2/1.0 PSI.

Figure 6.10 Model RB-1A, 3 ft^2 of filter felt area.

In performing the straining cycle oversize particles are collected in the felt, and a filter cake forms, which eventually causes a pressure drop increase and clogging of the felt.

It has been found that for bench-top testing, the use of felt cones (Figures 6.3 and 6.4) will predict quite well the results obtained in a given production performance. For example, place a given felt cone in a ring stand as shown, wet the felt with test liquid first, pour in test liquid, collect and evaluate level of straining.

The tie-on bag design (Figure 6.5 can, with reliability, operate up to 3 psi (10) psi with rounded bottom); however, the bag must be securely fastened to the feed pipe.

A more efficient and recommended use of the strainer felt medium is empolyed by the closure design of the Snap Ring bag, which will fit on and be supported by an adapter head. Note in Figures 6.6 and 6.7 that the adapter head has a ledge to support the metal closure ring sewn inside of the bag. Also, note that the felt selvage is placed on the outside of the bags. This is a precaution to avoid having two

Figure 6.11 Model RB-2A, installing the bags, 6 ft² of filter felt area.

layers of felt instead of only one layer "underneath" the metal closure ring of the Snap Ring bag. By tilting, the Snap Ring bag can be installed and seated on the support ledge of the adapter head (Figures 6.6 and 6.7). When pulled down by hand the bag is fully supported and sealed suitable for full-scale operation up to 15 psi.

The use of the Snap Ring bag provides complete usage of the strainer medium, rapid and easy installation and removal. It is automatically sealed, requiring no hose clamp or other tightening devices or are other precautionary steps needed. The Snap Ring bag design compared to the tie-on method is more convenient, reliable and reproducible, involving less labor. The various open systems methods are convenient and inexpensive to use.

CLOSED SYSTEM DESIGN DETAIL AND OPERATION

The advantages of continuous piping, handling solvents, good housekeeping and

Figure 6.12 Model RB-2A2L, 12 ft² of filter felt area.

the permanence of hardware require the use of a closed system. Felt bags employed as high flowrate strainers permit the design of compact equipment as pictured in Figures 6.9—6.20, the RB Hardware.

Felt Sealing

The Snap Ring bag in the open system employs a support ledge located on the inner periphery with respect to the bag (see Figure 6.6). The same Snap Ring bag in the closed system employed in the RB hardware design has a support ledge located on the outer periphery with respect to the bag (Figure 6.16).

In both cases the closure ring device is a convenience that causes the bag to spread out, available for complete straining performance. However, a significant detail is the location of the felt selvage. In Figure 6.7 for the open system, the selvage has been purposefully placed outside. In Figure 6.16 the selvage again has been purposefully placed, but inside. In both cases the sealing mechanism gives more reliable performance when using one uniform layer of felt. The selvage can be placed either inside or outside a given bag by pushing the felt inside out. The felts are symmetrical, being unaffected by direction of flow.

Figure 6.13 Model RB-4A, 12 ft² of filter felt area.

O-Ring Design

Referring to Figure 6.16, item nos. 8 and 9, O-rings supply the sealing method for the restrainer basket, felt bag and vessel. Item no. 7, another O-ring, provides the sealing for the feed line.

An occasional leakage problem can occur due to O-ring leakage as caused by uneven tightening of the bolts. The theory or mechanism of O-ring sealing is that about 2/3 of the cross-sectional diameter is contained by snug fit in a rectangular groove. As the opposing flange surface uniformly presses on the O-ring it fills the rectangular groove with enough excess elastomer to perform the primary sealing job.

The RB hardware restrainer basket, Item No. 3 Figure 6.16 has a double O-ring design to simultaneously seal the pressure vessel shell and the cover plate; a third O-ring seals the inlet. The RB hardware counts heavily on O-ring sealing which, in fact, gives reliable sealing and presents a cleanable surface on the restrainer basket flange, a desirable feature at the point of bag changing.

Figure 6.14 Model RB-4A2L, 24 ft² of filter felt area.

RB Hardware

The RB hardware is capital equipment with piping fixed and permanent. There is complete access to the inside of the vessel for inspection or cleaning. In operation the inside of the RB vessel is usually clean because all collected dirt is retained in the felt bag. This is in strong contrast to other filtration equipment where the vessels are usually very dirty.

Figure 6.16, the exploded view of Model RB-1A, indicates the elements of construction. To assemble an RB-1A one places the restrainer basket, No. 3, into the pressure vessel, No. 1. The cover, No. 2, is attached to the vessel by means of the hinge pin, No. 10. With the O-rings, nos. 7, 8 and 9 in place and installing the felt bag, No. 13, into the restrainer basket, the vessel is ready to bolt down nos. 5 & 6.

Model RB-1A is in effect, the basic module for the equipment line. The RB-1A equipment, 3 ft² of filter area, is expanded to be twice as long and it is called Model No. RB-1A2L, 6 ft² of filter area (Figure 6.17). If the RB-1A is expanded to have twice as many bags, it is called Model RB-2A (Figure 6.18) which includes twice the length RB-2A2L. See Table 6.4 for the complete listing.

Chapter 6. Arthur C. Wrotnowski

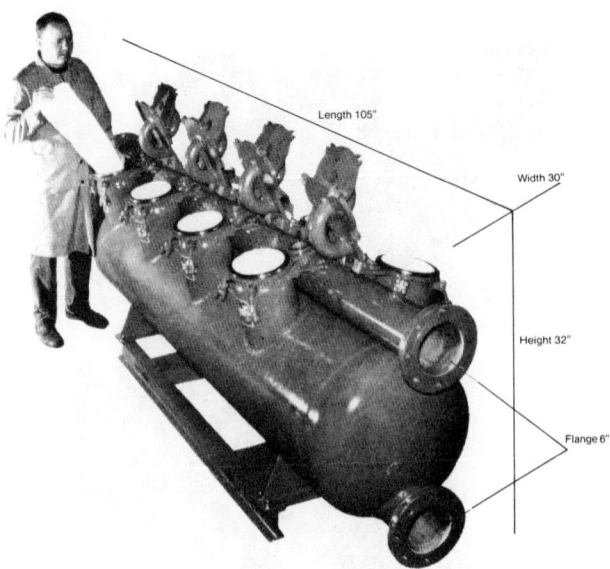

New Model GAFLO® Pressure Vessel Filter System — The new Model RB-8A filter bag/pressure filter systems, combines the filtering efficiency, fast easy installation, and convenience of the successful SNAP-RING® filter bag with the safety, high operating pressures, and other advantages of a closed system. Using eight No. 2 size SNAP-RING filter bags, the Model RB-8A filter has a maximum recommended flow rate of 1000 gpm.

Easy to install and operate, this new system permits significant savings in time, labor, and replacement costs while at the same time assuring top filtration performance.

GAF's performance-proven line of pressure filters is characterized by a simple design which consists of three basic components; a specially designed pressure vessel, SNAP-RING filter bags, and filter bag restrainers.

The pressure vessel is available in either mild steel or type 316 stainless steel with maximum operating pressure differentials up to 150 psi (10.5 kg/cm^2), ASME-coded to 150 psi (10.5 kg/cm^2) or 300 psi (21.1 kg/cm^2), the vessel conforms to ASME specifications as required by OSHA. The SNAP-RING filter bag is made of nonblinding synthetic fiber felt media (viscose, nylon, or polypropylene) and is characterized by high flow-rates and low pressure drop.

Figure 6.15 Model RB-8A2L, 48 ft^2 of filter felt area.

Model RB-X (Figure 6.9) is a 1 ft^2 filter area model, which is an inline model useful for modest-size applications. The small bag and equipment have the same advantageous equipment principles as the larger versions.

ASME Code

For ASME specification compliance hydrostatic testing 225 psi is applied to the restrainer basket (Figure 6.16 no. 3); also, independently, 300 psi is applied to the vessel (Figure 6.16 no. 1 itself).

Process Equipment Series Volume 1

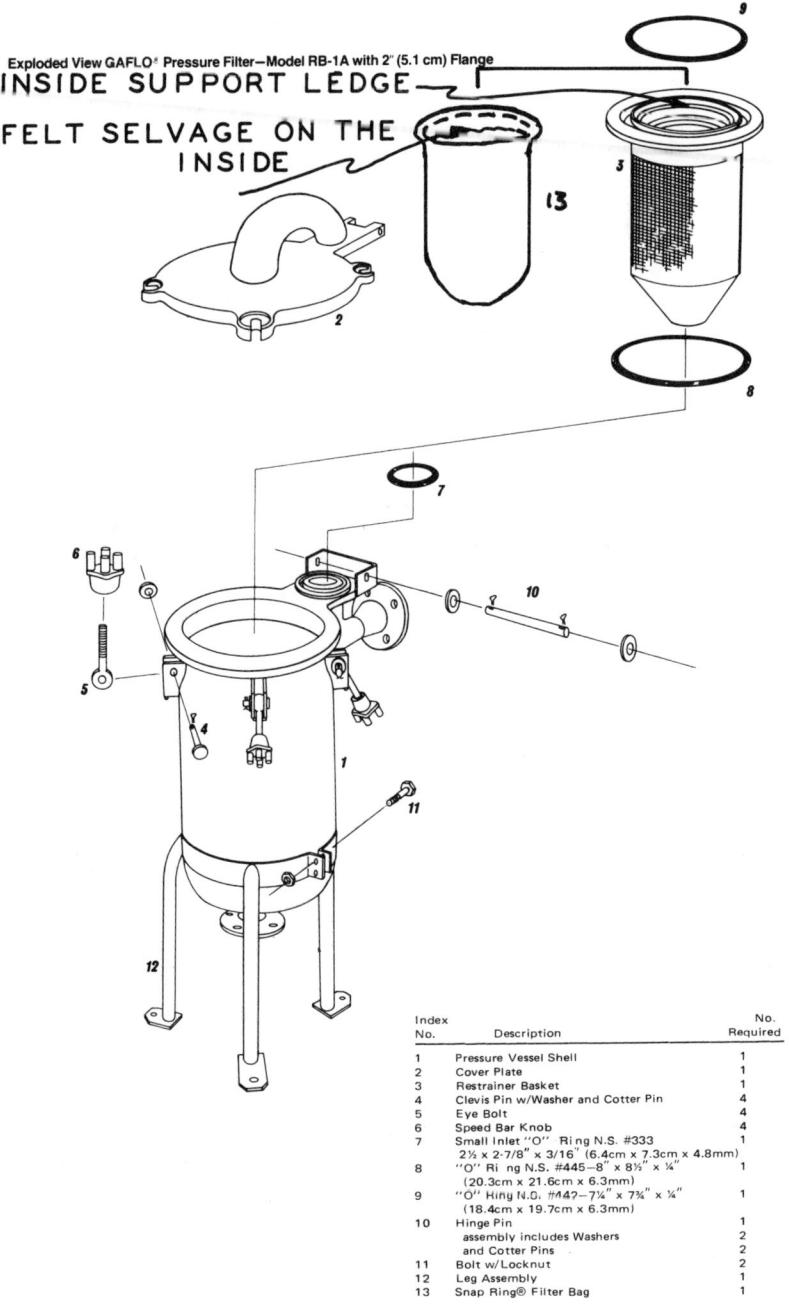

Index No.	Description	No. Required
1	Pressure Vessel Shell	1
2	Cover Plate	1
3	Restrainer Basket	1
4	Clevis Pin w/Washer and Cotter Pin	4
5	Eye Bolt	4
6	Speed Bar Knob	4
7	Small Inlet "O" Ring N.S. #333 2½ x 2-7/8" x 3/16" (6.4cm x 7.3cm x 4.8mm)	1
8	"O" Ring N.S. #445—8" x 8½" x ¼" (20.3cm x 21.6cm x 6.3mm)	1
9	"O" Ring N.S. #442—7¼" x 7¾" x ¼" (18.4cm x 19.7cm x 6.3mm)	1
10	Hinge Pin assembly includes Washers and Cotter Pins	1 2 2
11	Bolt w/Locknut	2
12	Leg Assembly	1
13	Snap Ring® Filter Bag	1

Figure 6.16 Exploded view of model RB-1A.

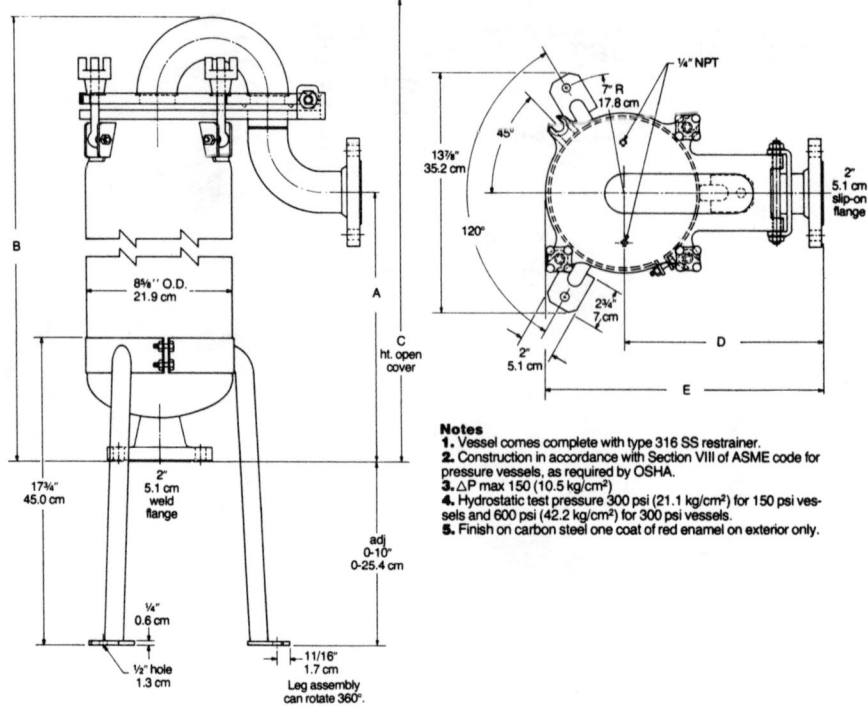

Gaflo Pressure Filter	Working Pressure		Weight		Dimension									
					A		B		C		D		E	
	psi	kg/cm²	lbs	kg	in	cm	in	cm	in	cm	in	cm	in	cm
RB-1A Mild Steel	150	10.6	100	45	15-1/4	38.7	25-15/16	65.9	33-7/8	86.0	12	30.5	16-5/8	42.2
RB-1A Mild Steel	300	21.1	120	55	15-3/4	40.0	27-1/16	68.7	34-3/8	87.3	12-5/16	31.3	16-15/16	43.0
RB-1A 316 Stainless	150	10.6	93	42	15-1/4	38.7	25-15/16	65.9	33-7/8	86.0	12	30.5	16-5/8	42.2
RB-1A 316 Stainless	300	21.1	113	51	15-3/4	40.0	27-1/16	68.7	34-3/8	87.3	12-5/16	31.3	16-15/16	43.0
RB-1A2L Mild Steel	150	10.6	120	55	29	73.7	39-11/16	100.8	47-5/8	120.9	12	30.5	16-5/8	42.2
RB-1A2L Mild Steel	300	21.1	140	64	29-1/2	74.9	40-13/16	103.6	48-1/8	122.1	12-5/16	31.3	16-15/16	43.0
RB-1A2L 316 Stainless	150	10.6	113	51	29	73.7	39-11/16	100.8	47-5/8	120.9	12	30.5	16-5/8	42.2
RB-1A2L 316 Stainless	300	21.1	133	60	29-1/2	74.9	40-13/16	103.6	48-1/8	122.1	12-5/16	31.3	16-15/16	43.0

Figure 6.17. Dimensions and models of RB-1A and RB-1A2L.

Operation Procedure of RB-1A

1. Select felt bag by micrometer rating or felt cone testing.
2. Assure RB-1A components are present, *i.e.,* the restrainer basket and O-rings.
3. Install bag.
4. Close lid, bolt down uniformly.
5. Start pump and recirculate.
6. Fill vessel slowly and vent air.
7. Employ full flow.
8. Test for particle removal efficiency.
9. Proceed until flowrate decreases to economic minimum.
10. Stop pump, open vessel, remove bag, repeat cycle.

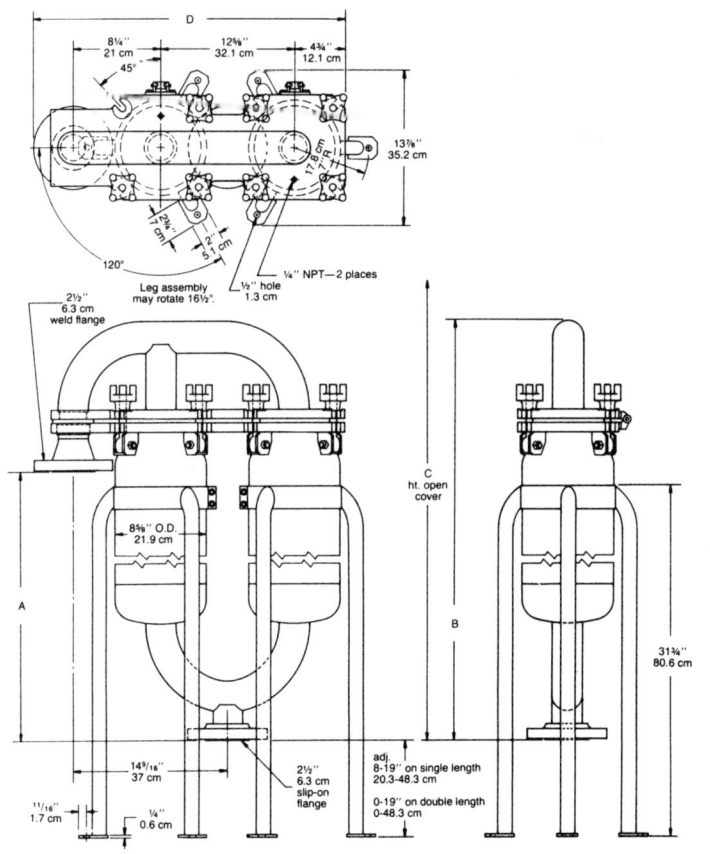

Gaflo Pressure Filter	Working Pressure		Weight		Dimension							
					A		B		C		D	
	psi	kg/cm²	lbs	kg	in	cm	in	cm	in	cm	in	cm
RB-2A Mild Steel	150	10.6	200	91	24-7/8	62.1	37-13/16	96.0	38-7/8	98.7	29-1/8	74.0
RB-2A2L Mild Steel	150	10.6	246	112	38-3/16	97.0	51-9/16	131.0	52-5/8	132.7	29-1/8	74.0
RB-2A Mild Steel	300	21.1	246	112	24-7/16	62.1	38-1/16	96.7	38-15/16	97.5	29-3/8	74.6
RB-2A2L Mild Steel	300	21.1	285	129	38-3/16	97.0	51-13/16	131.6	52-11/16	133.8	29-3/8	74.6
RB-2A 316 Stainless	150	10.6	190	86	24-7/16	62.1	37-13/16	96.0	38-7/8	98.7	29-1/8	74.0
RB-2A2L 316 Stainless	150	10.6	236	107	38-3/16	97.0	51-9/16	131.0	52-5/8	132.7	29-1/8	74.0
RB-2A 316 Stainless	300	21.1	236	107	24-7/16	62.1	38	96.5	38-15/16	97.5	29-3/8	74.6
RB-2A2L 316 Stainless	300	21.1	276	125	38-3/16	97.0	51-3/4	131.4	52-11/16	133.8	29-3/8	74.6

Figure 6.18 Dimensions and models of RB-2A and RB-2A2L.

SELECTING THE FELT STRAINER BAG MEDIUM

Without previous felt straining experience it is best to perform a bench top test employing a felt cone. The cone can be placed in the support ring of a laboratory stand as indicated in Figure 6.4. The applicable control test can then be performed. There is, of course, a large variety of particles and liquids in this area of interest.

Chapter 6. Arthur C. Wrotnowski

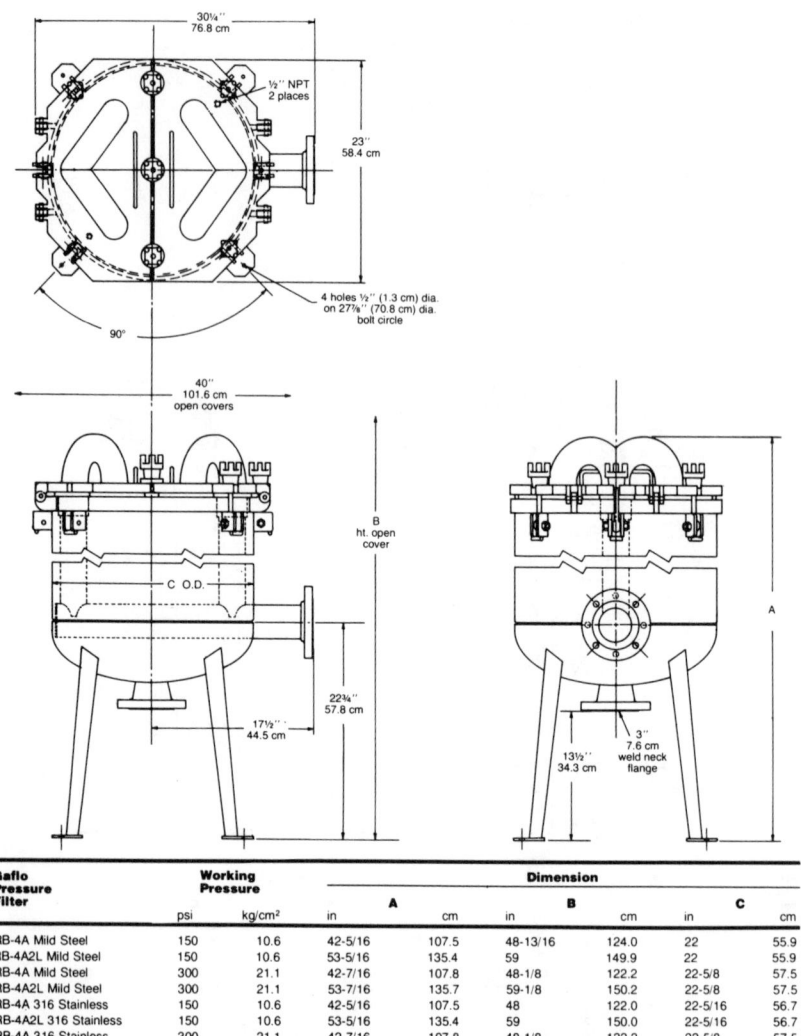

Gaflo Pressure Filter	Working Pressure		Dimension					
	psi	kg/cm²	A in	cm	B in	cm	C in	cm
RB-4A Mild Steel	150	10.6	42-5/16	107.5	48-13/16	124.0	22	55.9
RB-4A2L Mild Steel	150	10.6	53-5/16	135.4	59	149.9	22	55.9
RB-4A Mild Steel	300	21.1	42-7/16	107.8	48-1/8	122.2	22-5/8	57.5
RB-4A2L Mild Steel	300	21.1	53-7/16	135.7	59-1/8	150.2	22-5/8	57.5
RB-4A 316 Stainless	150	10.6	42-5/16	107.5	48	122.0	22-5/16	56.7
RB-4A2L 316 Stainless	150	10.6	53-5/16	135.4	59	150.0	22-5/16	56.7
RB-4A 316 Stainless	300	21.1	42-7/16	107.8	48-1/8	122.2	22-5/8	57.5
RB-4A2L 316 Stainless	300	21.1	53-7/16	135.7	59-1/8	150.7	22-5/8	57.5

Figure 6.19   Dimensions and models of RB-4A and RB-4A2L.

For example, molten wax, dispersed wax, natural gums and varnishes, glue, latex and paint, when being manufactured, require particles of some nature to be removed to achieve product and quality control. Whether straining delicate flocculent materials, gelatinous particles or high-density hard solids, oversized particles can usually be strained out by calibrated felt strainer bags.

Process Equipment Series Volume 1

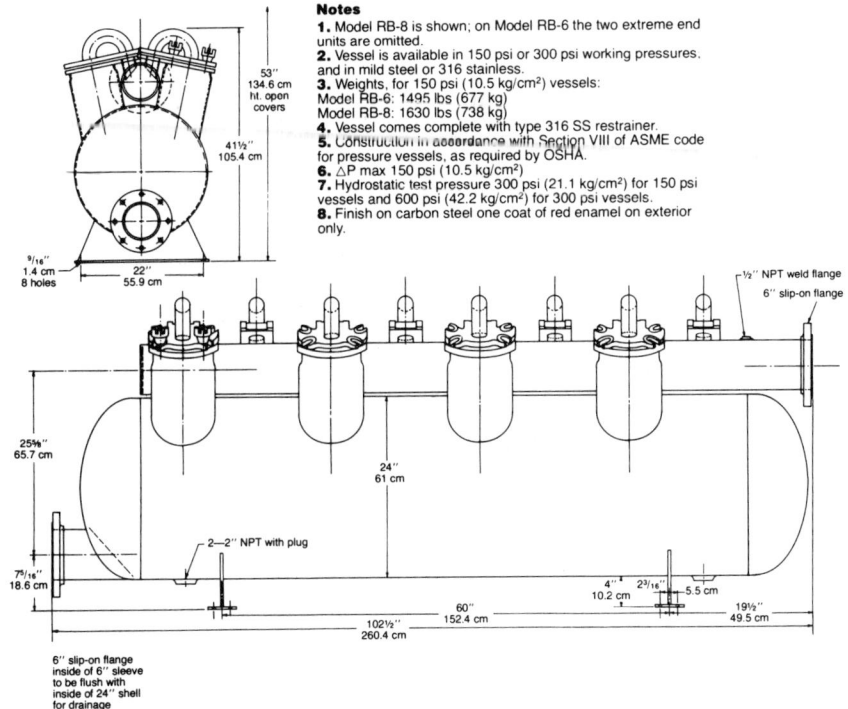

Figure 6.20 Dimensions and models of RB-6 and RB-8.

When selecting a felt medium the investigator should consider the nature of the particle and the process objective. For example, the particle size ratings given in Table 6.2 were determined by means of a flyash slurry. Flyash particles are hard vitreous siliceous spheres, which form the slag from burning powdered coal in power plants.

The particle size classification action will probably be somewhat different when handling some other irregularly shaped or gelatinous-type particle. However, it is expected that an optimum strainer felt can be found in the range of sequential pore-sized felts as given in Table 6.2.

The Felt Strainer Bag Series A-G of Table 6.2 basically perform oversize particle removal at some given level from the $1-200\mu$ range. These families are based on the synthetic fibers available as needed to meet the general needs of the chemical process industries. The many synthetic fiber diameters are useful. In felt straining the mechanism variable is principally fiber diameter rather than felt density.

The strainer bag basic construction has a fleece or working staple fiber interlocked onto a strength member, an open scrim. The open scrim permits fine particle passage. It is a tough industrial fabric suitable for general rough use. The use of the scrim does not interfere with particle separation yet provides strength for industrial applications.

193

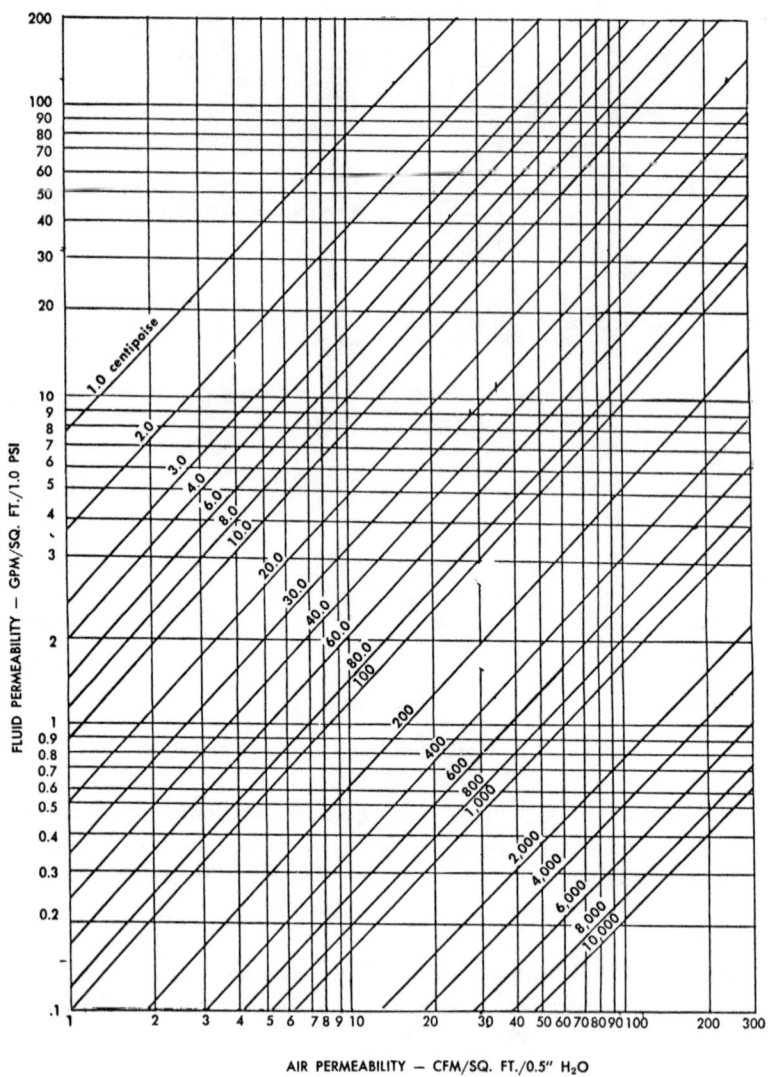

Figure 6.21 Pressure drop — felt strainer media, air permeability, liquid permeability at various viscosities.

Discussion of Table 6.2

The rayon viscose series is a general purpose "workhorse" line. It is supported by a strong nylon scrim. Note that both nylon and viscose have similar chemical stability properties; for example, stable in caustic, stable in aliphatic and aromatic

Table 6.2. Felt Strainer Bag Media — Air Permeability, Nominal Values

	Particle Size Rating (μm)	Air Permeability (cfm/ft^2/12-in·H$_2$O)	Application
A.	Rayon, Viscose (Nylon Scrim)		
	5	50	Solvents, paints,
	10	100	general use
	15	150	
	25	250	
	50	300	
	100	350	
	200	400	
B.	Polypropylene		
	1	15	Acid, Alkali,
	5	50	general use
	10	100	
	25	150	
	50	200	
	100	300	
C.	Nylon		
	5	50	Heavy duty,
	10	100	alkali and solvents
	25	150	
	50	200	
	100	300	
D.	Nomex		
	5	25	Elevated temperature
	25	150	
	50	250	
	100	350	
E.	Teflon		
	30	50	Extreme corrosion and temperature
F.	Wool-Silk		
	1	10	Fine particle retention
G.	Wool-Cotton		
	3	20	Fine particle retention

Note: No. 1 See Table 8 for Metric Air Permeability Units.

solvents and both have histories of being FDA amenable for use with foodstuff. The rayon-nylon series is not suitable in acid or biologically active environments.

The polypropylene strainer bags are especially useful in very corrosive acid or alkali chemical exposures except hot aromatic solvents. Polypropylene is com-

Chapter 6. Arthur C. Wrotnowski

Table 6.3. Chemical and Thermal Component Compatabilities — Ten Examples

No.	Liquid Involved	Felt Strainer Medium	RB Vessel	O-Ring	Closure Ring
		Fiber (Code-Rating*)			
1	Raw Latex	RV-200	Steel	Neoprene	Cadmium (CD) Plated Steel
2	Clay Slip	RV-25	Steel	Neoprene	CD-Steel
3	Gums & Waxes	RV-25	Steel	Neoprene	CD-Steel
4	Varnish (hot)	NY-25	Steel	Viton	CD-Steel
5	Red Rouge Fe_2O_3	WS-1	316SS	Viton	CD-Steel
6	Metallic Paint (Aluminum Flake)	RV-50	Steel	Neoprene	CD-Steel
7	Hot Melt Adhesive	HT-25	Steel	Silicone	CD-Steel
8	Sulfuric Acid, 10%	PO-5	PVC-Coated	Neoprene	316SS
9	Caustic, 10%	PO-5	Steel	Neoprene	316SS
10	Enamel	RV-25	Steel	Neoprene	CD-Steel

Note:

Chemical Compatability**

Fiber	Fiber Code	Operating Temperature (°F)	Acid	Alkali	Solvent
Rayon Viscose	RV	180	Over 5% — poor	Excellent	Excellent
Nylon	NY	220	Poor	Excellent	Excellent
Polypropylene	PO	240	Excellent	Excellent	Aromatics—poor
Nomex	HT	350	Good	Good	Excellent
Wool silk	WS	180	Poor	Poor	Excellent
Teflon	TE	500	Excellent	Excellent	Excellent

*Rating is given micrometers   **For detail values consult fiber manufacturers

pletely stable in biological environments and is excellent for general purpose use. The nylon strainer bags are tough and thermally resistant especially in anhydrous and oxygen free environments. Nomex can be used for hot melt adhesives at 350°F or for other high-temperature uses. A special strainer bag of Teflon®* is useful for exotic chemicals; wool-silk and wool-cotton are useful for extra fine straining, principally in solvent or aqueous slurries.

SIZING THE EQUIPMENT

Since the bag system is essentially a batch operation, the size of the equipment determines the frequency of bag changing. However, it is quite possible to get an effective continuous system by using two vessels placed in parallel, each vessel having enough capacity to handle the full flow and solids encountered. While one

---

*Registered trademark of E. I. du Pont de Nemours and Company, Inc., Wilmington, Delaware.

Table 6.4. Pressure Vessel, RB Hardware ASME Coded for 150 psi

Item No.[a]	Identification Model No.	Strainer Medium Area (ft^2)	Felt Strainer Bags Employed		Reference Figure No.
			Quantity	Size	
1	RB-X	1	1	RB-X	9
2	RB-1A	3	1	1	10
3	RB-1A2L	6	1	2	—
4	RB-2A	6	2	1	11
5	RB-2A2L	12	2	2	12
6	RB-4A	12	4	1	13
7	RB-4A2L	24	4	2	14
8	RB-6A2L	32	6	2	—
9	RB-8A2L	48	8	2	15

[a] Item Nos. 3, 5 and 7 can be considered the "stretch" version of Item Nos. 2, 4 and 6 respectively.

[b] felt strainger bag sizes, effective straining area.

Size No.	Area (ft^2)
RB-X	1.
1	3.
2	6.

bag vessel is in use the other can have new bags installed; meanwhile the flow is continuous.

After selecting a felt strainer bag medium, as dictated by chemical compatability and particle size rating, and determining whether a batch or continuous method will be used, the size of the equipment to be used, and be analyzed by:

1. the pressure drop, based on liquid flowrate and viscosity; and
2. solids to be removed.

In general, and for practical reasons, we are considering large quantities of liquids with relatively small quantities of oversized particles to be removed. By determining the pressure drop under these conditions, we essentially size the equipment based on the relative power cost.

Closed Filter Bag System

The example given in Table 6.5 has been calculated for all of the larger hardware given in Table 6.4. It will be seen that the pressure drop developed by bag and vessel is quite measureable. The pressure drop developed by the solids collection is the main controlling variable for long life.

As a textile material is is convenient to classify the strainer felts by air permeability (6). This value can be converted to various permeabilities as shown in Figure 6.21. As a convenience, an approximate conversion formula applying to 12 various viscosity units is given in Table 6.6.

Chapter 6. Arthur C. Wrotnowski

Table 6.5. Procedure for Determining the Clean Felt Pressure Drop in RB-Hardware (Example: Pressure Drop Calculated for Eight Different Pieces of Equipment

Problem: Determine the clean felt and hardware pressure drop for a pigmented neutral water solution thickened with cellulosic sizing.

Given:
1. Desired Flowrate, gpm   200
2. Liquid Viscosity, centipoise   440
3. Desired Level of Cleanliness, *i.e.,* particle size μm   25
4. Liquid Specific Gravity   1.2

Procedure

Step No. 1. Select strainer bag medium.
From Table 6.3 Rayon Viscose is chemically compatible.
From Table 6.2 the 25-μm Rayon Viscose medium has an air permeability of 250 cfm/ft² /½ in. H_2O.
Step No. 2. Convert Felt air permeability to liquid permeability.
From Figure 6.21 at 440 Centipoise, the felt medium liquid permeability is 9 gpm/ft² /1.0 psi.
Step No. 3. Hardware pressure drop (ΔP_H).
Consider the RB hardware as follows:

	Figure No.	Pressure Drop (ΔP_H), psi 440 Centipoise	
		100 gpm	200 gpm
RB-1A	23	6.5	13
RB-1A2L	23	6.5	13
RB-2A	23	4.5	9
RB-2A2L	23	4.5	9
RB-4A	23	2.8	5.6
RB-4A2L	23	2.8	5.6
RB-6A2L	Estimated	1.8	3.6
RB-8A2L	Estimated	1.0	2.0

Step No. 4. Felt pressure drop (ΔP_F)

$$\Delta P_F = \frac{FR}{LP \times A}$$

where FR = desired equipment flowrate
       LP = clean felt liquid permeability (Step No. 2 above), gpm/ft² /1.0 psi
       A = felt area (Table 6.4) ft²

Step No. 5. Total pressure drop (ΔP), psi.

$$\Delta P = \Delta P_F + \Delta P_H \quad SP.GR.$$

where SP.GR = specific gravity of liquid

Model								ΔP
RB-1A	=	$\frac{200}{9 \times 3}$	+	13		1.2	=	24.5
RB-1A2L	=	$\frac{200}{9 \times 6}$	+	13		1.2	=	20.0
RB-2A	=	$\frac{200}{9 \times 6}$	+	9		1.2	=	15.2

(Continued)

Table 6.5. Procedure for Determining the Clean Felt Pressure Drop in Hardware (Continued)

Model								ΔP
RB-2A2L	=	$\dfrac{200}{9 \times 12}$	+	9		1.2	=	13.0
RB-4A	=	$\dfrac{200}{9 \times 12}$	+	5.6		1.2	=	8.9
RB-4A2L	=	$\dfrac{200}{9 \times 24}$	+	5.6		1.2	=	7.8
RB-6A2L	=	$\dfrac{200}{9 \times 36}$	+	3.6		1.2	=	5.1
RB-8A2L	=	$\dfrac{200}{9 \times 48}$	+	2.0		1.2	=	3.0

Comment on Solids Capacity. The average dirt-holding capacity for a given size No. 1 (3 ft^2) strainer bag on ordinary loam, rust scale or otherwise granular-type contaminant is 5 lb. However, the clogging and blinding properties of an individual filter cake will dominate the final "close off."

(Concluded)

A summary of the procedural steps in Table 6.5 follow:
1. Convert viscosity units to centipoise.
2. Determine liquid permeability of the felt medium.
3. Determine pressure drop in RB hardware.
4. Calculate the total pressure drop considering total felt area and the individual models.

The total pressure drop results are clean felt and hardware values. From the design flowrate and calculated pressure drop, the energy consumption pumping costs can be calculated for economic comparison, which represent long-time straining no solids buildup.

From a practical point of view the average dirt holding capacity for a No. 1 size bag (3 ft^2) for ordinary, loam, rust scale or otherwise granular type dirt is 5 lb. The potential dirt removal from a given feed stream is a variable that must be determined experimentally, yet this factor will determine the straining life of the bag, frequency of changing bags and is the major factor in determining the size of the equipment.

Economic Comparison

Therefore, in sizing the equipment the investigator can compare the capital expenditure and the associated rate of depreciation against: (1) the increase in power cost of high-pressure equipment; and (2) the increase in labor cost resulting from bag-changing frequency. (In some cases there may be no labor increase over a fixed labor expense.).

The maximum rate of depreciation for chemical processing equipment is nine years, based on the Internal Revenue Code Section No. 167 using the Asset Depreciation Range System as of 1978.

Table 6.6. Converting Various Viscosity Units to Centipoise (for subsequent use on Figure 6.21)

1. Viscosity Units at 70° F

Units	Fabtor	Units	Factor
Demmler #1	14.6	Parlin Cup #15	98.2
Demmler #10	146.	Redwood admiralty	10.87
Engler	34.5	Redwood standard	1.095
Ford Cup #4	17.4	Saybolt furol	10.
MacMichael	1.92	Saybolt seconds universal	1.0
Parlin Cup #2	187.0	Stormer	Approx. 13

2. Conversion Formula[a]

$$\text{Absolute Viscosity (Centipoise)} = \frac{(\text{Viscosity Units}) \times \text{Factor} \times \text{SP} \cdot \text{GR}}{4.62}$$

[a] The conversion formula is approximately valid when the numerator is greater than 250.

Range of Materials Strained or "Scalped"

- Paint, Lacquer, Varnish & Inks
- High Viscosity, Cellulosic, Alginic, Carageenin & Xanthate
- Raw Latex, Dispersions of Synthetic Resins & Polymers
- Hydrocarbons & Aromatic Solvents
- Reflective Pigments; Pearl Essence & Metallic Paint
- Dyestuff Pastes & Pigments
- Sandmill "Take-off" Disperse Dyestuff
- Abrasives, Magnetic Tape Particles
- Bio-mass
- Ceramic Clay Slip
- Monomers, Resin, Glue
- Pectin, Gelatin, Milk

Open Filter Bag System

The open bag system has a variety of combinations that may be used in inexpensive arrangements, yet quality straining performance can be expected.

As described above the main controlling factor is the amount of solids collected in each case and the decision for the open system is to select a functional system of adequate solids capacity. Pressure drop can be calculated in each case from Figure 6.21; however, the solids collected determine the capacity.

ALLIED EQUIPMENT AND PIPING ARRANGEMENTS

Both open and closed felt bag straining systems require a minimum amount of external pumps, tanks and piping. The felt straining process is essentially a low-pressure drop operation and, therefore, low horsepower motors are employed. It is not unusual to employ gravity feed. Suction and discharge piping and tanks are, of course, required.

The selection of pumps is mainly a function of the liquid being pumped as opposed to the effect on the felt strainer operation. High-viscosity liquids, such as paint and varnish, require positive displacement pumps. Felt straining by means of positive displacement pumping proceeds normally. Felt strainer bags have been found to be functional with all types of pumps, including those that have oscillatory pressure.

The RB hardware is quite amenable for adaptation as a portable pump and filter rig. The number of applications are as wide as the number of possible liquids. For example, portable rigs are practical for paint, lube oil, wire drawing oil, lithographic ink, felt pen nib ink formulation, etc.

The amount of control equipment used to evaluate straining is numerous. Some examples follow:

1. Hegman Grind Gage — paints
2. Millipore septum — general
3. Andreason pipette — solids settling method
4. Metal screens — general

Open System Piping

For open systems the piping in effect terminates at the bag, as shown in Figure 6.7 and 6.8, where the straining itself occurs. Manifolding, for example, indicates that many piping variations are possible. Some practical applications include strainer bag discharge into: (1) an electroplating bath; (2) a paint or chemical dip; (3) straining of water supply for immediate use; or (4) ink straining discharge directly into a fiber drum for shipment.

Closed System Piping

The RB hardware is used as a conventional filter. Figure 6.22 has four piping arrangements; recycle, normal run, air evacuation and evacuation by pumping, which are conventional operations.

It is quite practical to employ various lineups, such as two vessels in series, the first removing a coarser fraction to promote greater capacity in a second operation. It is also practical to make a continuous system placing two vessels in parallel. When one vessel handles the flow and straining, the second vessel would have a new bag installed to be ready to take the load as needed.

EQUIPMENT AND FELT INSPECTION

The use of strainer bags is deceptively simple. The use of a common felt bag tends to disarm one technically. However, it is now well established that felt

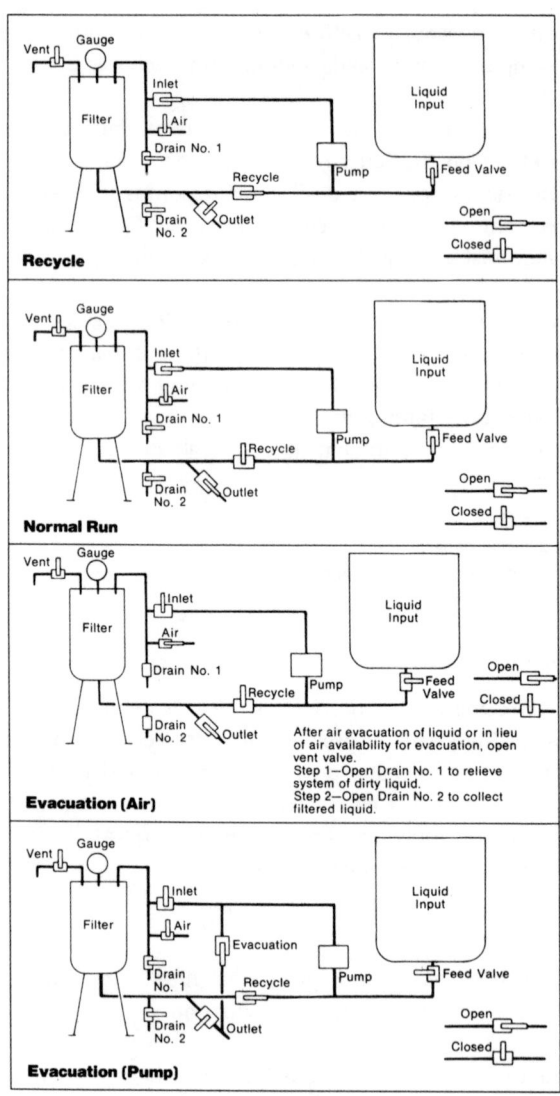

Figure 6.22 Operation cycles for RB-1A.

straining technology makes a contribution to the particle size classification field and has a place in the chemical process industry. To achieve reproducible and reliable felt straining a number of details must be under control.

With regard to the felt bag, there are visual judgment inspections to make as follows:

1. Quality sewing job should include correct dimensions, uniform seams, etc.

2. Uniform felt fabric.
3. Proper fiber content is to be determined by straining and air permeability tests.
4. In a given felt bag place the felt selvage opposite the "support ledge" (Figures 6.6, 6.7 and 6.16) as discussed under Closed System Design Detail and Operation, p.184. An uneven selvage will interfere with snap-ring sealing. It is more reliable to have one uniform layer of felt at the point of sealing.

With regard to the equipment inspection, a brief product control checkoff list is given in Table 6.7. The subjects covered are intended to assure good workmanship and performance reliability.

In addition to this section and Table 6.7 following the text is a formal and complete specification itemizing the proper design of the subject equipment, design and performance specification.

CHEMICAL AND THERMAL STABILITY

Today's chemical processing industry (CPI), when using the felt strainer bag system, challenges the engineer to balance off the choice of materials available against the chemical environment challenging the equipment. In actual fact, the RB hardware, having few components, simplifies material selection.

The various strainer bag components with comments follow:

No.	Description	Comment
1.	O-rings	A large choice of elastomers are available. Stability to temperature, corrosiveness and solvents is usually possible.
2.	Felt	A large range of textile fibers, mainly stable to chemicals, are available (see Table 6.2)
3.	Vessel and closure ring device	Steel, 316 stainless steel and coatings (fluorocarbon, PVC, etc.).
4.	Adaptor head	Delrin, celcon, polypropylene and 316 stainless steel.

The corrosiveness of CPI liquids fall into a few categories:

1. *Benign:* water, latex dispersions, natural gums, botanicals.
2. *Degradation attack:* acid, alkali, reducing.
3. *Solubility attack:* solvents.
4. *Extremely corrosive:* exotic.

All materials of construction have some chemical weakness that can degrade and break down; however, there are so many choices available chemically and thermally stable materials of construction are usually available.

Table 6.7. Product Control Checkoff List for Model No. RB-1A Vessel (Figure 6.16)

1. Top flange of vessel, Item No. 1, and mating cover, Item No. 2, must be flat and parallel. Sealing surfaces must be free of scratches, dents and weld splatter.
2. A straight edge placed on the restrainer basket, Item No. 3, and the inlet must be parallel to the sealing surfaces of Item Nos. 1 and 2, above.
3. The two weld neck flanges (inlet and outlet) must be square and parallel, respectively, to the sealing surface of Item No. 1.
4. Vessel opening dimension must be a minimum of 7.156 in.
5. With the restrainer basket, Item No. 3, in place and the cover plate, Item No. 2, closed, the hinge pin, Item No. 10, must not bind.
6. The restrainer basket, Item No. 3, flange thickness is to be 0.325 in. in thickness.
7. Swing eye bolts are to be aligned properly to match slots and counterbore.
8. Angle of cover, when open, should be between 115° and 125°.

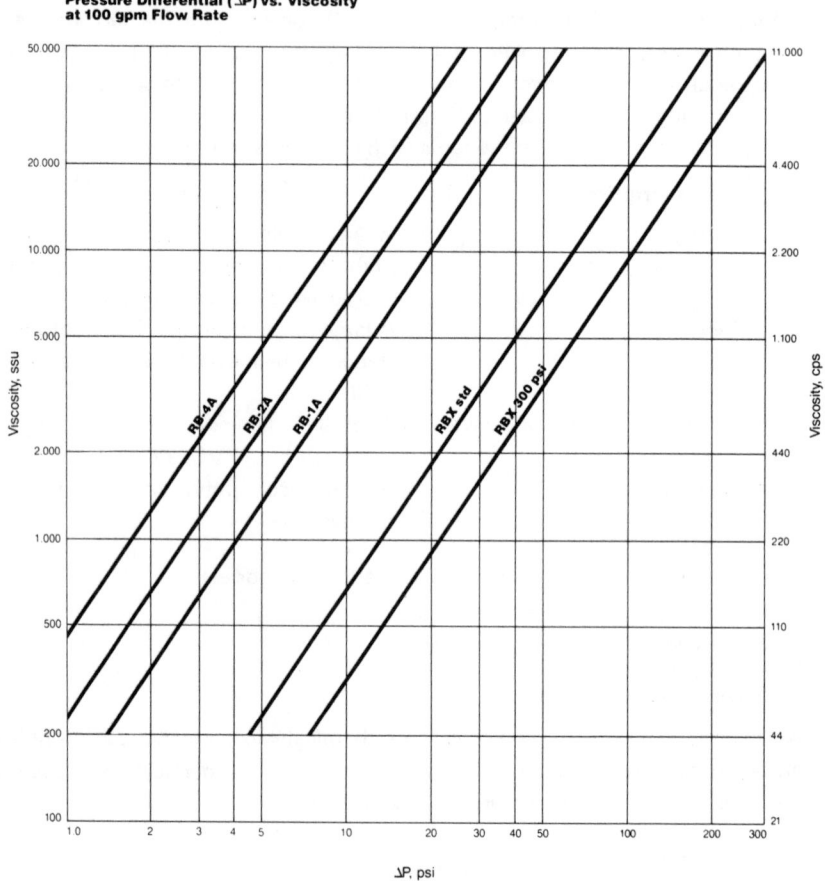

Figure 6.23 Pressure drop — RB hardware at various viscosities.

Table 6.8. Conversion English to Metric Flowrate Units (Ref. No. 3).

Liquid Flow Rate	
1.0 gpm/sq. ft./1.0 psi	= 580 liter per sq. meter/min1.0 kg. per sq. cm.
1.0 liter per sq. meter/min /1.0 kg. per sq. cm.	= 0.00173 gpm/sq.ft./1.0 psi

Air Flow Rate	
1.0 cfm/sq.ft./½" H_2O	= 305. liter per sq. meter/min./124 Pascal (Pa.)
1.0 liter per sq. meter/min./124 Pa	= 0.00328 cfm/sq.ft./½" H_2O

TROUBLE SHOOTING

Both the open and closed strainer bag systems are quite reliable mechanically. However, there can be recurring failures that will be avoided by correct usage as follows:

Closed System   **Action**

1. <u>O-Ring Leakage</u>
   — Check to see if O-rings are properly installed
   — Bolt down cover plate to be parallel to the pressure vessel flange.

2. <u>No Flow</u>
   — Air bound — vent air.
   — Excess gelatinous particles.
   — Piping system defective.

3. Closure Ring Device Leakage
   — Check closure ring alignment; fit and overlap seam compressibility.

4. <u>Bag Burst</u>
   — Check bag size and fit in the restrainer basket; spread out bag when installing

5. <u>Bag Degraded</u>
   — Wrong bag selection, *e.g.*,
   (1) Use of polypropylene in hot Xylene.
   (2) Use of viscose in biologically active conditions.

Open System

1. <u>Unsupported Bag Breaks Under Pressure</u>
   — Replace bag at lower pressure, *e.g.*, 10–12 psi. Average ultimate strength is 18 psi;

2. <u>Bag Breaks When Supported By Basket</u>
   — Check method of installation, *e.g.*,
   (1) Place Snap-Ring bag on adaptor; pull sharply to set the ring
   (2) Place restrainer over the bag and support the restrainer on the adaptor properly
   (3) Start operation

Chapter 6 Arthur C. Wrotnowski

DESIGN AND PERFORMANCE SPECIFICATION

Introduction

The felt strainer bag system has been developed for liquid straining to a high performance level and has achieved general acceptance as an engineering tool. The strainer bag system is designed to replace and augment the filter cartridge and screens, based on superior design and performance.

Important features of the strainer bag system are as follows:

1. *Economic Comparison.* One bag has the equivalent capacity of several cartridges.
2. *Disposal of Filter Cake.* Collected contaminate is contained in the bag.
3. *Filtrate Cleanliness.* Contaminate does not reach the downstream side of the strainer bag.
4. *Labor Expense.* Minimizes downtime.
5. *Time Elapsed Per Cycle.* High flowrates.
6. *Pump Wear.* Low pressure drop.
7. *Power Cost.* Less energy consumed in processing.
8. *Warehouse Expense.* Less inventory requirement in units and space.
9. *Minimum Operator Exposure Time.* Possible due to rapid bag change time.
10. *Pressure Leaf Filters and Plate and Frame Filters.* Roughing filters have been successfully replaced.
11. *200,000 CPS liquids.* Strained on a practical basis.
12. *Above 200 μ.* Woven mesh bags in the 150–800 μ range are economical and practical.

The above highlights result in fewer manufacturing, processing and inventory costs and are associated with higher production efficiency. The strainer bag system is recognized as one of the simplest and most effective straining devices available. The standards and design criteria are hereby established for a complete strainer bag system.

Manufacturers of filter bag systems should meet all of the design requirements. When systems fall short of meeting the basic requirements full performance of the felt strainer medium cannot be obtained.

The sophistication in the design features that has resulted in its simplicity and effectiveness is easily overlooked. There are 10 pertinent design features to consider. A detailed itemization follows on the 10 pertinent design considerations to aid in understanding and selecting a strainer bag system.

Contaminate contained in the Bag

1. All contaminate should be collected and contained in the strainer bag.
2. A flat cover with a long radius is employed to direct the flow of the incoming liquid from the inlet pipe to the center of the strainer bag.
3. This is to ensure low pressure drop, an even flow pattern and equal distribution of liquid in the bag.
4. A uniform particle distribution and cake collection will result throughout the run.

Positive 360° Circumferential Sealing of the Bag
1. The opening or mouth of the strainer bag should be completely contained and compressed around the total 360° circumference to ensure no bypassing of unfiltered material.
2. A machined seat for the filter bag's closure ring is located in the top entrance of the restrainer basket flange. Both the diameter and depth of the seat and held to ± 0.005 in. tolerance.
3. Closing the cover and tightening down for operation assures that an equal compression force is applied to the bag's closure ring around its entire 360° circumference. This ensures that, in operation, no bypassing of the filter bag with unfiltered liquid can occur.

Quick and Complete Access to the Bag
1. Access to the filter bag should be quick and easy without interrupting or breaking connections in the pipeline.
2. On loosening and swinging down the hold down bolts of the vessel, the cover can be raised and backed against a simple hinge arrangement. This action completely exposes the internals of the vessel for ease of handling.
3. The hinge supports the cover in the open position at a 20–30° backward tilting angle. This prevents damage to the vessel and components as well as ensuring operator safety in handling.
4. The angle at which the cover is held open prevents accidental closing of the cover even if knocked or bumped while in the open position.

No Contaminate Reaches the Downstream Side
1. *No contaminate* should be allowed to pass to the downstream side of the bag during operation or upon removal of spent bags — whether or not a blowdown is employed.
2. The above is an extremely important feature of a properly designed strainer bag system. To positively assure that no contaminate passes to the downstream side of the strainer, even on removal of a spent bag, requires positive assurance that no contaminate can form above or outside of the strainer bag during operation. The complete O-Ring and flange design creates this important and necessary feature (see sections entitled "Positive 360° Circumferential Sealing of the Bag;" "Positive Vessel Sealing under all Operating Conditions;" and "Fixed Sealing Devices.")
3. Even if a blowdown is used, contaminate that has been allowed to collect above or outside of the bag will inevitably reach the downstream side of the bag on removal of the spent filter bag.

No Residual Contaminate Buildup
1. The vessel should be designed with smooth, clean, internal lines and surfaces so that no residual contaminate can be trapped or allowed to build in the vessel.
2. Use flange connections, smooth long radius bends and necks down the vessel body at the bag inlet (when possible) so that there are no shelves or broken areas for contaminate buildup.

3. Clean design of the vessel and strainer bags is necessary to obtain USDA acceptance.
4. The clean smooth continuous wetted surface areas allows the standard vessels to be easily coated or lined for improved chemical resistance or release properties.

Positive Vessel Sealing Under All Operating Conditions

1. Positive sealing of the vessel at the inlet section and the vessel to atmosphere under the entire spectrum of chemical exposures, pressures and viscosities, is a necessity.
2. A machine-raised seat containing an O-Ring is utilized to seal the incoming liquid as it passes from the inlet line into the long radius return bend located on top of the vessel falt cover.
3. A machined flange to which the restrainer basket is affixed contains the O-Rings for sealing the vessel to atmosphere. Both the inlet raised seat and the basket flange have the same machined height and are located on a flat ring plate fixed to the top of the vessel body.
4. The equal machine heights of both the inlet seat and basket flange act as stop bars so that proper predetermined compression is applied to the O-Rings for positive sealing.
5. O-Rings are selected for seals because of their maximum sealing efficiency with a minimum of applied force.
6. A complete selection of elastomers in standard size O-Rings is readily available to meet the variety of chemical exposures.
7. To ready the vessel for operation the holddown bolts are swung into locating slots in the cover and the bar knobs are turned down to apply pressure.
8. The locating slots in the cover are counterbored to the diameter of the bar knob seat so that they cannot twist out or cock when a turning tightening force is applied.
9. In the open position, when the holddown bolts are swung downward at the side of the vessel body, they are held away from contacting the vessel by a stop device designed into the bolt's clevis bracket. This is an operator safety feature and prevents chipping and damage to both vessel body and holddown bolt assembly.

Fixed Sealing Devices

1. O-Rings, gaskets or seals must be locked in place and easily cleaned even when the vessel is inoperative and opened. Sealing materials should not be allowed to pull free to drop onto the floor or into the vessel even when opening the vessel after handling "tacky" and high-viscosity liquids.
2. Most of the details and some of the design features have been previously discussed in both 1. and 5. in the previous section, concerning the raised inlet O-ring seat and restrainer basket flange. A press-fit, O-ring groove design in each of these devices positively contains the O-ring without damage and still does not

allow the O-ring to pull out of the groove under severe operating conditions.
3. It is not necessary to remove the O-ring to clean the groove as no liquid should get by the exposed face of the O-ring. Therefore, one merely wipes the exposed surface to ensure complete cleaning.
4. Longer than normal O-ring life can be expected using the press fit groove design. The grooves are machined to a fixed depth and width to require a slight press fit of the O-ring in the groove. The annular space provided in the groove is sufficient to allow the O-ring to deform under compression without damage, and provide more than its normal resistance to compression for more effective sealing.

Restrainers Should Not Impair Flow or Life
1. The restrainer basket should not impair the flowrate or onstream life of the strainer filter bag media from its full potential.
2. The function of the restrainer basket is to completely support the strainer bag medium so that lateral forces cannot be applied to the medium. This ensures uniform pore size and straining results throughout the run, regardless of the pressure drop buildup.
3. A standard restrainer basket for general specification compliance in all vessels is constructed with 10 X 10, 0.047 in diameter, #316 stainless steel wire mesh.
4. Each 150-psi restrainer basket is inspected and pressure tested to 225 psi before installing it in a vessel.
5. There is no impairment in strainer bag flow or life when employing a woven metal wire restrainer.
6. Perforated metal having ¼ in. holes allows the medium to stretch, blocks off over 50% of free area and can potentially unload at high pressure drop. The perforated metal causes sideward flow through the media, increases pressure drops and a loss in media life, up to 25% of the media's potential.

Easily Cleanable
1. The vessel and components should be easily cleanable for changeover of products.
2. The cleanability features of vessels were covered in detail in the section entitled "No Residual Contaminate Buildup."

Meet ASME and OSHA Requirements
1. The vessel should meet the ASME code requirements under Section VIII of the unfired pressure vessel code along with OSHA regulations.
2. An acceptable standard line of vessels is ASME coded, meets OSHA requirements and is USDA approved.
3. There are a few states that have requirements beyond Section VIII of the ASME unfired pressure vessel code in that they require vessels to be designed and tested to 2 times working pressure rather than the 1½ times that the code requires.
4. It is good practice to standardize all vessels to the 2-times design and test pressure requirement in both the 150-psi and 300-psi working pressure models. The temperature design limits within the 2-times design and test pressure is —40

to +600°F for carbon steel; and −60 to +300°F for the #316 stainless steel models.
5. Standard strainer bag pressure vessels have approximately 1/8 in. corrosion allowance built in above the standard ASME requirements.

REFERENCES

1. Grace, H. P. "The Art and Science of Liquid Filtration," paper presented at the Jubilee Meeting of AIChE Philadelphia, PA, June 25, 1958.
2. Harris, C. C. "Some Aspects of Screen Sizing," *Col. Eng. Quart.* 15 (1): 18–23, 48 (1961).
3. Lauterbach, H. *Textile Res. J.* 25: 143 (1955).
4. Wrotnowski, A. C. "Felt Straining Technology," *Filtration Separation* 13 (5) (September/October 1976).
5. Wrotnowski, A. C. "Final Filtration With Felt Bag Strainers," presented at the AIChE 84th National Meeting Atlanta, GA, March 1, 1978, available on Microfiche as Paper No. 54C.
6. ASTM D461-67 "Testing Felt."

CHAPTER 7

INCLINED PLATE SETTLERS

INGOLF V. JANERUS
Parkson Corp.
Fort Lauderdale, FL

INTRODUCTION

Hazen presented his surface loading theory in 1904. This theory suggests that the capacity of a settling tank is independent of its depth and only dependent on its surface area. Thus, the necessary background for shallow depth sedimentation has been present a long time. Although several inclined plate settler designs were disclosed early, only in the last 10 years has this compact technique been seriously introduced.

Presently it is estimated that approximately 1,000 inclined plate settlers are in use worldwide. Approximately half of these are located in North America. The rapid growth in popularity is mainly due to design improvements, which have substantially improved the performance and thus resulted in even more compact and economical settlers. These often can be delivered as package units ready for installation and can be moved from one location to another. The space requirement for an inclined plate settler is often 10% or less of the land area needed for a conventional settler. The key design improvements are related to flow distribution and introduction of feed between the plates. These design improvements originated from Sweden (1) where most of the work was done to correctly apply inclined plate settling to a number of applications. Today there are inclined plate settlers operating in North America on applications spanning from potable water treatment to wastewater treatment, including clarification and thickening of many process streams. Prediction of the performance of a specific plate settler is not as straight/forward as for a conventional settling tank. Therefore, extensive pilot testing has been necessary to properly apply inclined plate settlers to various applications. The author is fortunate enough to be able to draw from the experience of close to 100 pilot tests performed in the U.S. as well as more than 400 existing installations.

Inclined plate settlers vary in design from tube settlers, which are mainly used to upgrade existing clarifiers, to advanced settlers like the Lamella^{TM} Gravity Settler (Figure 7.1). The sizes of the installations vary from small package units with about 10 m^2 (100 ft^2) of settling area to large concrete basin installations with more than 100 m^2 (10,000 ft^2) of settling area.

Chapter 7. Ingolf V. Janerus

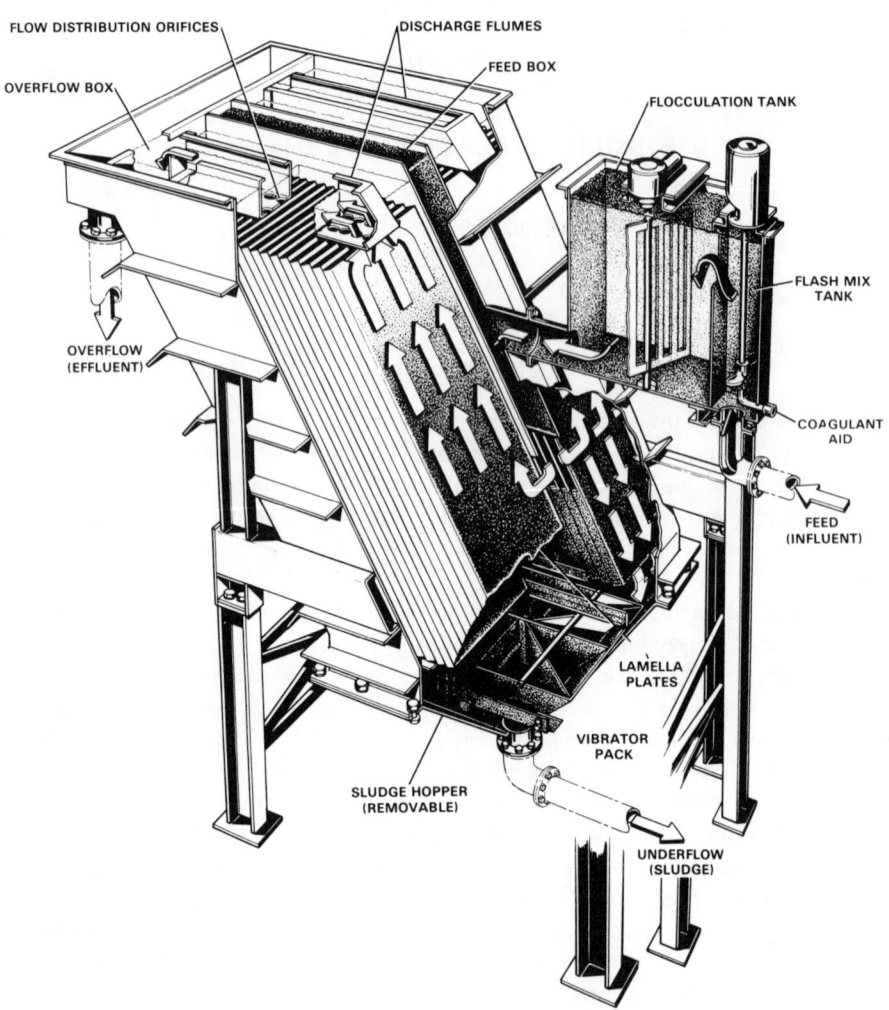

Figure 7.1 Inclined plate settler (Lamella Gravity Settler). Note that the feed enters the plates from the side and the effluent is throttled to ascertain even distribution of the flow between the plates.

THEORY OF INCLINED PLATE SETTLER

The basic equations for sizing settling basins were formulated more than 70 years ago. Consider for the moment an ideal settling basin (Figure 7.2). The suspension enters at one end of the basin, flows uniformly along its length at velocity V_L and exits at the other end. The particles settle towards the bottom at velocity V_S. The trajectory of the particles is indicated by the vector, V_P. If this trajectory takes the particles to the bottom of the basin before they reach the far end, it is assumed that they are removed from the liquid. Therefore, a particle starting at the top must settle through the distance, H, at velocity, V_S, in the same time (or less) that the liquid is in the basin. Thus:

$$\frac{H(m)}{V_s(m/hr)} \leq \frac{L \times W \times H(m^3)}{Q(m^3/hr)}$$

Simplifying,

$$\frac{Q}{L \times W} = \frac{Q(m^3/hr)}{A(m^2)} \leq V_s(m/hr)$$

where A is the settling area of the basin and Q/A is known as the "overflow rate" or "surface loading" (and is usually expressed in m/hr, gpm/ft^2 or gpd/ft^2). From this relationship it can be seen that all particles are removed that have a settling rate equal to or greater than the overflow rate, and that the height (or detention time) of the basin is not one of the main parameters effecting the separation efficiency.

The fact that this is true can be illustrated another way. Compare Figures 7.2 and 7.3, the only difference being that the second basin is half the height of the first. As a result, the detention time is only half as much and the suspension moves

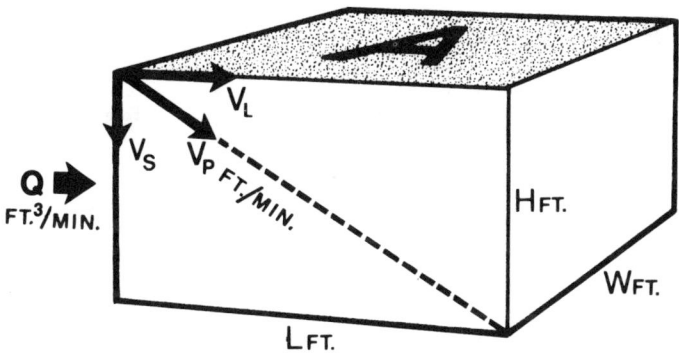

Figure 7.2

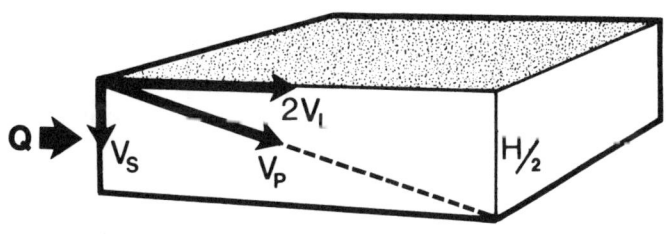

Figure 7.3

through the basin at twice the velocity, V_L. The trajectory of the particles has only half the slope, but since the basin is only half as deep, the particles are still removed.

If the height of the basin is reduced to a few inches and a number of such units are stacked on top of each other, the result is a primitive shallow-depth sedimentation device. Figure 7.4 shows a unit containing 10 parallel compartments. Theoretically, it can handle 10 times the flowrate as could the same basin without any plates. The liquid detention time is one-tenth as long. However, the same separation efficiency is achieved since the overflow rate is still the same ($10Q/10A = Q/A$). Note that the settling area now includes the area of all the plates.

The shallow depth sedimentation device shown in Figure 7.4 is impractical since it is difficult to remove sludge from the plates. Either the space between the plates must be large enough to accommodate mechanical scrapers or the unit must be shut down periodically and backflushed. Both systems are used occasionally for special applications, but in general they are impractical.

For a more general solution, the plates are inclined so the sludge will be self-draining. Figure 7.5 shows an arrangement containing 10 plates set at an angle of 60° above the horizontal. The total settling area is derived as above, but the plate area must be multiplied by the cosine of the angle to correctly determine the

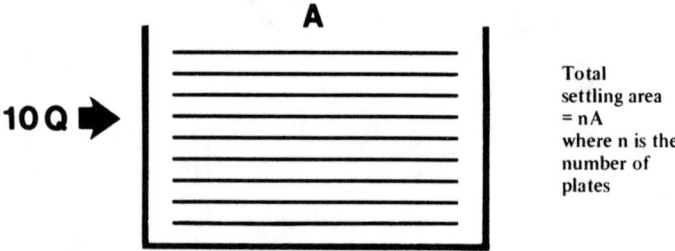

Figure 7.4 Total settling area = n x A, where n is the number of plates.

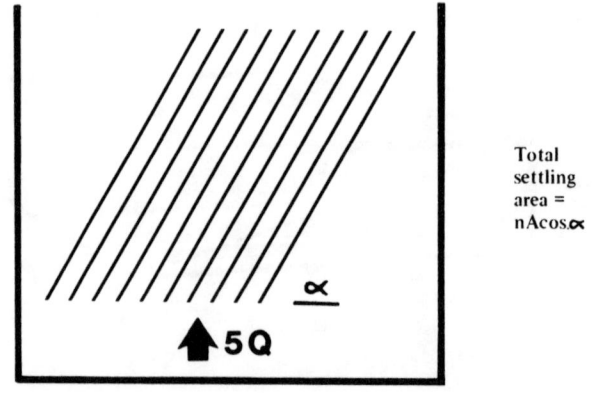

Figure 7.5 Total settling area = n x A cos (X).

capacity and overflow rate. (Thus only the projected area of each plate on a horizontal plane is "counted.") In this example, the total settling area is 10A cos. $60° = 5A$ and the capacity of the unit is 5Q.

The design shown in Figure 7.5 is exemplified by the present-day tube settler. Tubes are used instead of plates but the basic principle is the same. The tubes are generally about two inches square in cross section and two feet long. This design, however, suffers from two very serious limitations. First, there is no means provided to ensure that the flow is uniformly distributed throughout the settler. As a result, parts of the settler may be overloaded while other parts may be underloaded. Secondly, the sludge that collects in the tubes must settle through the incoming feed to reach the bottom of the basin and be removed. As a result, the solids may be reentrained by the feed. Due to these limitations, the settler must be operated at a loading rate of only 25–50% of its theoretical "overflow rate" (2, 3). Additionally, tube settlers and similar devices have no provision for hindered settling.

The main advantage of the inclined plate settler is its compactness. A measure of compactness is the amount of projected surface area "stacked" per unit land area.

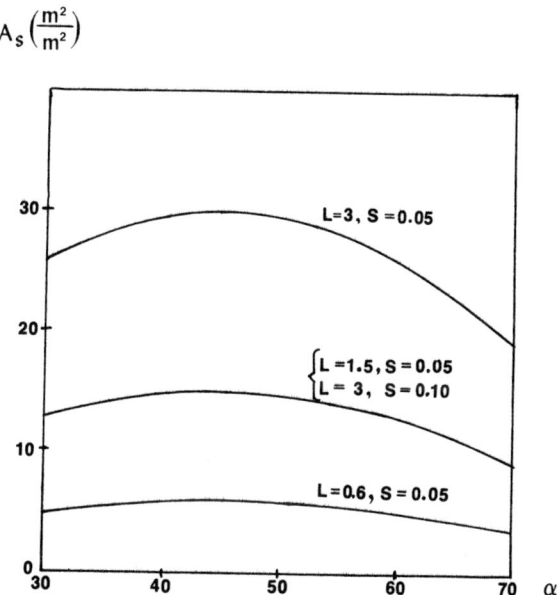

Figure 7.6. The compactness, A_s (projected area/land area) as a function of the angle of inclination, α, with the plate length, L, and the perpendicular distance between the plates, s, as parameters.

The variables of importance for compactness are: (1) perpendicular distance between plates, (2) plate inclination, and (3) plate length.

As will be discussed later there are limitations for all variables and Figure 7.6

should only be used as an illustration. It should also be noted that Figure 7.6 is a theoretical compactness assuming an infinite number of plates. In practice, the number of plates will be finite and space for introduction of feed has to be included. Therefore, the specific surface for a practical design is generally 35—65% of the theoretical value. From Figure 7.6, however, it is clear that for maximum compactness the plate spacing and inclination should be as small as possible while the plates should be as long as possible. Figure 7.6 also shows that the amount of projected area easily can be made to exceed 10 times the land area occupied.

EQUIPMENT DESIGN

It was mentioned earlier that the basic principles for inclined plate settler design have been known for a long time but that the technology was not commercially exploited until the last 10 years. Lately, compactness has been more and more appreciated. To enhance compactness it is important that each inclined plate (or rather the spacing between two plates) be utilized at its maximum capability. Recent (patented) designs have accomplished that by ascertaining an even distribution of flow to each plate spacing and by preventing the settled solids from mixing with the feed as the solids leave the plates and drop into the sludge compartment. From Figure 7.6 it is clear that we also should use as small a plate spacing and as long plates as are practical. A discussion of these and other important parameters follows.

Flow Distribution

To use the plate area efficiently the flow has to be distributed evenly between different plate spacings as well as widthwise across the plate spacing.

Flow Distribution Between Plate Spacings

If no means for flow distribution are included in the design the flowrate (velocity) through some plate spacings will be much larger than the average (100% larger is not uncommon). For such a piece of hardware one often has to select the design conditions for the whole unit to correspond to the plate spacings with the highest throughput or close to it. Thus, *provided all other things are equal,* a total of 50—100% extra projected area would have to be supplied for a design without flow distribution means compared to a design with efficient flow distribution means.

There are two reasons for possible maldistribution:

1. The velocities in the feed and effluent chambers are, at least in some points, considerably higher than the average velocity through the plate spacings. This means that the passage through the plate spacings doesn't offer the dominating pressure drop in the equipment and that it thus "doesn't matter" which way the liquid chooses through the plate pack.
2. The second reason is that an installed unit is never perfectly levelled. Thus, one flow path through the unit might offer a larger static head than another path and consequently different size flowrates will flow along the different paths.

Maldistribution can be avoided if a sufficiently large pressure drop is connected with the passage of the liquid through each and every plate spacing, *e.g.,* consider a design according to Figure 7, which includes an orifice on the effluent side of each plate spacing. If the orifices are sized to produce a 50-mm (2 in.) pressure drop then a 6-mm (¼ in.) level difference between two different points in the plate settler would still only produce a 6% difference in flowrate. Generally a pressure drop of 50—75 mm (2—3 in.) is recommended, which can be accomplished by including one or two orifices of 12—25 mm (½—1 in.) per plate spacing.

There are principally two ways of assuring that the passage of each and every plate spacing is associated with the same pressure drop:

1. Some earlier designs positioned the pressure drop-producing restriction such that the liquid had to pass through it to enter the specific plate spacing. Each plate spacing has its own identical restriction. This is not a practical solution as the high velocity in the restriction will destroy all flocs. Even for unflocculated material it is a bad design as the high-velocity jet will destroy the laminar flow for a large part of the plate spacing.

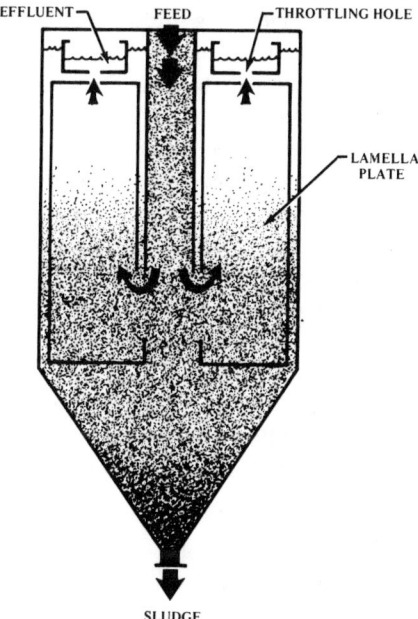

Figure 7.7 Lamella Gravity Settler (end-view).

2. Position the restriction at the exit from each plate spacing. The restrictions are now positioned in the clear phase, which also will keep them from plugging. The most desirable design is to use orifices as restrictions. This design is patent protected. A less desirable alternative is to allow the effluent from each individual plate spacing to exit over a weir.

Flow Distribution Across the Width of the Plate

One single plate (spacing) is best utilized when the upward liquid velocity is the same across the width of the plate. To accomplish that three criteria have to be fulfilled:

1. The entrance velocity has to be kept low otherwise jet streams will develop and travel up the plate. This excludes influent throttling as throttling always is connected with high velocities.
2. The length of the plate has to be long compared to the width of the plate for a stable and even flow pattern to develop. A length/width ratio of at least 5:1 is preferred.
3. This liquid should be removed from the top of the plate as evenly as possible, *i.e.*, one drawoff point should serve as small a width of the plate as possible. A side effluent weir serves the whole width of the plate which may create undesirable flow patterns.

Feed Entry

It has already been mentioned that the plate area will not be used efficiently if the solids dropping from the plates into the sludge compartment are mixed into the feed, which then enters the plates from below. This can be avoided by first introducing the feed into a feedbox or feed channel (Figure 7.1 and 7.7) from which it has access to the plates through large openings in the sidewalls of the plates (feed-ports). The bottom of the feed channel should be kept open so that all solids settling out will fall down into the sludge compartment. If the bottom is closed or almost closed buildup will occur for most applications, which will lead to plugged openings into the plate area, malperformance and excessive maintenance.

Once the feed has entered into the plate area as few obstructions as possible should be present. Flow distribution means between the plates, especially in the sludge-rich lower parts of the plates, will obstruct transportation of solids and cause pluggage for most applications.

The introduction of the feed into the feed channels as well as the flow in the feed channels have to be such that flocs are not sheared and that no "jetting" occur that can disturb the flow through the feedports into each plate spacing.

Plate Spacing, Length and Width

The solids that settle out are sliding down the plate surface and dropping into the sludge hopper. The fluid flow must be laminar — turbulent flow will resuspend the solids that are sliding down the plate. At a given surface loading rate the flowrate between two plates is proportional to the length of the plates (as the surface area is proportional to the length). Thus, for very long plates, the velocity would be high enough to create turbulent flow. For most applications 3-m (10 ft)-long plates will offer a compact enough solution without creating excessive velocities between the plates.

The width of each plate — or flow channel if the plate is divided into compartments — has to be moderate compared to the length, otherwise the flow profile will be far from the desired plug flow. The wider the plate is, the larger is the risk that jet streams will develop, which of course means that the plate area is badly utilized. As mentioned before, the length/width ratio is recommended to be at least 5 to 1, preferably larger.

The compactness of the plate settler is greatly influenced by the plate spacing, e.g., choosing 25-mm (1 in.) plate spacing instead of 50-mm (2 in.) will double the projected area that can be installed in a given tank. For most applications, however, a spacing of less than 2 in. will encourage bridging of solids between the plates. The result is excessive shutdowns for cleaning of the equipment.

Plate Inclination

The plates are inclined to facilitate sludge removal from the plates. To make the sludge move down the plate, the component of the gravity force along the plate has to be larger than the forces resisting the movement. One of the two resisting forces is the friction force between the sludge particles and a layer of sludge, whichever is less. The other resisting force is the shear force between the sludge layer and the bulk of liquid flowing up the plate.

For sludges such as metal hydroxides, with a density close to the density of water, the net gravity force is fairly small and a steep inclination is necessary. An inclination of 55° has been found to be sufficient to ensure safe sludge transportation. For denser sludges 45° is usually sufficient. As a plate with 45° inclination has a projected area that is 23% larger than the projected area at 55° inclination, it is obvious that the 45° inclination is more economical where it can be used safely.

Sludge Withdrawal

The compactness of the plate settlers results in that up to 550 m^2 (6000 ft^2) can be supplied as preassembled "package units" The simplest means of sludge collection and withdrawal system is a hopper as shown in Figure 7.1.

A shallow hopper will almost certainly rat-hole. In a rat-holing hopper the velocity is much higher in the central parts of the hopper than along the hopper walls. The result is that dilute sludge is breaking through into the discharge and sometimes that sludge actually clings to the hopper corners creating deposits that may increase in size and thus decrease the available hopper volume, as well as blocking sludge release from the plates. This, of course, will result in malperformance both regarding overflow quality and sludge concentration.

One manufacturer includes a low-frequency vibrator in the hopper section. The vibrations aid sludge concentration as well as transportation. The latter because the vibrations are lowering the apparent viscosity of the sludge, which increases its mobility. Even with the vibrator the sludge hoppers are made with a side angle of at least 55° toward the horizontal.

The relatively short detention time for the sludge in the hopper is a handicap in applications where high sludge concentrations are created by compression during fairly long times. For finely dispersed mineral type sludges it has been demonstrated that low amplitude vibrations have a positive effect on compression. The vibrations enhance packing of the sludge particles (Figure 7.8) (1). Especially for larger installations it is common that plates are mounted above a basin containing a rake mechanism. For further discussions regarding underflow concentrations please see the following sections.

This section has discussed a number of parameters that determine how well the available projected area will be used. The reader should keep in mind that this varies greatly between different designs.

SIZING METHODS. SCALE-UP FROM PILOT TESTING

The most common way of sizing gravity settlers is by making column settling tests as described in a previous chapter. Very good predictions of settler performance can be made this way provided that the samples tested are representative. The end result of the settling test can be looked on as the performance of an *ideal* settler as a function of the surface loading rate applied. (For hindered settling the solids loading rate will also enter.) Safety margins are added to allow for nonideal hydraulics and sometimes to allow for expected variations in settling properties.

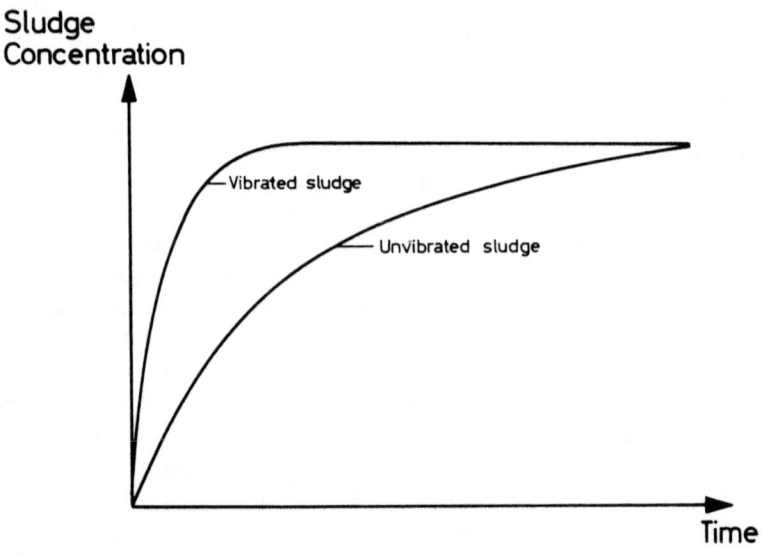

Figure 7.8 Effect of low amplitude vibrations on compression of sludge

For inclined plate settlers the column settling test is also the basic sizing tool. However, to simulate the shorter detention time in a plate settler the settling test is usually performed in a 500-ml graduated cylinder, from which the top 100 ml is withdrawn after a certain time to simulate a certain loading rate. The depth of the 100-ml layer corresponds roughly to the distance between two plates. As the settling time is comparatively short (generally 1–10 minutes), it is very important that the timing of the test is precise. The withdrawal of liquid has to be made rapidly but also in a manner that doesn't create any vertical currents in the cylinder. To perform good reproducible settling tests takes quite a bit of training.

The graduated cylinder test is more convenient than a long column test when operating at elevated temperatures as it is easier to thermally equilibrate a graduate, *e.g.,* in a constant water bath. Apart from this, however, it takes less training to perform good tests in a long column. Still graduated cylinder tests are recommended when sizing plate settlers because excessive natural flocculation is avoided and good mechanical flocculation is much easier to simulate in a graduate. The widespread use of polyelectrolytes eliminates the influence of natural flocculation for most applications today.

From the graduate settling tests we obtain a relationship between overflow clarity and surface loading rate, which, for a desired clarity and given flowrate, gives us the necessary projected area to be multiplied with a proper safety factor. The safety factor, however, depends on how efficiently a certain design utilizes the available area, which may lead to safety factors between 1.25 and 2.0 or above, depending on the design. To finalize, the equipment design experience will enter the picture as plate inclination, plate spacing, vibrations, etc. have to be determined. Other factors that have to be checked are wether the type of solids and/or the volume of sludge sliding down the plates will restrict the maximum loading rate that can be used without scouring of solids from the plates and/or developing inefficient settling conditions. Thus, the final equipment selection should always be left to a responsible supplier who can draw on his experience. For most applications it is also good practice to be extra conservative when sizing is based on one sample. This, of course, can be avoided by frequent sampling and by pilot testing.

Hindered Settling, Compression

For applications involving high sludge/overflow ratios, hindered settling and compression are often the phenomena determining the size of the thickener. For conventional equipment the sizing is often based on column tests, from which solids flux curves can be generated. The solids flux curve (4) gives a relationship between the solids loading rate (mass of solids per unit area) and underflow concentration. Compression depends on sludge bed height and detention time and can also be somewhat predicted from column tests.

For a theoretical model of hindered settling in a plate settler the reader is referred to the literature (5). Three zones can be distinguished between each pair of plates.

1. one lower zone closest to the lower plate surface consisting of concentrated material traveling down the plate surface;
2. one bulk zone in which hindered settling takes place; and
3. one upper zone closest to the upper plate surface consisting of clear liquid traveling up the plate as a jet stream.

The solids flux method is a tool which can be used — with proper modifications — also for inclined plate settlers, but the specific type of settler as well as other variables enter into the detail design and sizing of the equipment. Contact an equipment vendor with operating installations for the final design.

To reach the desired underflow concentration the sludge sometimes has to be subject to compression. The amount of compression is a function both of the depth of the sludge layer and the time the sludge is subject to compression. The compactness of the hopper of an inclined plate settler generally offers a short detention time. For many types of sludges (*e.g.* mineral type) this can be compensated for by applying low amplitude vibrations, as demonstrated by Figure 7.8. For other applications, the so-called LGST concept is a way of obtaining high sludge concentrations (inclined plates mounted above a thickener compartment with a picket fence rake mechanism.

Pilot Testing

Pilot plant testing is made to verify performance predicted from laboratory settling tests. Pilot tests should be made on applications for which there are no full-sized installations operating using the contemplated type of plate settler. It should be kept in mind, however, that a pilot test is a major undertaking as pretreatment of the feed etc. has to be simulated to correspond to the future full-sized system. As this often has to be made in a provisional way the pilot test will demand more supervision as manpower often has to be substituted for automatic controls.

The pilot test is made to verify the scale-up from laboratory testing. It is then obvious that the pilot equipment should simulate the full-sized equipment as well as possible. The critical dimensions should be the same as in full-sized equipment and, generally, the more plates a pilot unit contains, the better the simulation. As the flow distribution system is the most critical variable, the design engineer should especially evaluate whether the pilot unit is "idealized" in this respect compared to the full-sized system.

COMMON APPLICATIONS.
COMPARISON WITH CONVENTIONAL EQUIPMENT

Besides the chemical processing industry we find inclined plate settler installations in most industries where conventional settlers are used, *e.g.*, in iron & steel, metal finishing, mining, power, pulp & paper, potable water treatment and even biological wastewater treatment. This section will give examples of applications of interest to the CPI as well as pointing out the strong and weak points for inclined plate settlers compared to conventional settlers.

Comparison with Conventional Clarifiers/Thickeners

To start with, inclined plate settlers were generally installed in free settling applications where a space shortage was forcing the selection of a compact settler, *e.g.*, clarification of washwaters from a metal finishing shop. Development of more advanced plate settlers and extensive test work has expanded the area for which plate settlers can be used to cover almost all applications. For most applications the plate settlers represent the lowest cost alternative to do a certain job, but for a few applications for which the short detention time is a serious handicap from a process standpoint, the inclined plate settler will be uneconomical unless its compactness in itself offers great advantages one way or another. Below is listed a number of advantages and disadvantages of an inclined plate settler compared to a conventional settler. Most of them are direct functions of the compactness.

Advantages

1. Reduces land area needed by about 90%
2. Generally costs less on a total installed basis. The difference often gets dramatic when special materials of construction are required to withstand corrosive conditions, pressurizing etc.
3. Package units are factory preassembled resulting in minimal field erection work. May be installed indoors or covered at much less cost.
4. Is easy to insulate and has much lower evaporation and heat loss. Its low weight make it possible to install on high locations in buildings.
5. Contains a small inventory which is important if the inventory is valuable or if the settler needs to be emptied periodically.
6. Is not affected by wind or thermal currents to any practical extent
7. Can easily be integrated with the process equipment to be close to the source of the feed and can be moved from one location to another.
8. Has far fewer moving parts that require maintenance and that may malfunction.
9. Easy access to sludge piping aboveground.
10. The LGST concept can produce previously unknown sludge concentrations for certain applications.

Disadvantages

1. Small liquid detention time, which minimizes natural flocculation.
2. Small sludge inventory when thickener is used for storage of sludge. Can be compensated for by using the LGST concept or external storage tank.
3. Small sludge detention time for compression. Can be compensated for by vibrations for some sludges and by using the LGST concept.
4. Cannot be recommended for applications with large scaling potential, especially not where the scale cannot easily be removed.
5. Some types of solids will stick to the plate to a certain extent. This will cause periodic flushing of the plates. Severe cases of "stickiness" should be avoided.

During the last decade the use of synthetic polyelectrolytes has become widespread. This means that for many applications efficient flocculation can take place in one to two minutes detention time, resulting in very compact flocculator/settler combinations as shown in Figure 7.1. Mechanical flocculators are to prefer compared to baffle-type flocculators as the latter are less efficient per unit volume and the efficiency also greatly depending on the flowrate. The use of polyelectrolytes also means that there are very few situations today where natural flocculation is of importance.

Materials of Construction

The tank is commonly made of carbon steel, epoxy painted or with some special coating. Stainless steel is also common as well as rubberlined carbon steel. Smaller tanks are sometimes made of fiberglass-reinforced plastics (FRP). The small dimensions of the inclined plate settler often make special and expensive materials feasible.

Different kinds of plastics are the most popular choices for plate materials. Different grades of FRP and polyvinylchloride (PVC) are the most commonly used plastics while stainless steel is the preferred metal.

It should be pointed out that the different plate settlers are not only different concerning hydraulic design but also regarding mechanical integrity etc., which affects the reliability and useful life of the equipment.

Example of Applications in the CPI

There are inclined plate settlers in operation on a variety of applications in the CPI. They vary from production of process water to being an integral part of the process system and to product recovery and/or wastewater treatment.

Production of process water and/or potable water has already been discussed in the section about the LGST. For smaller flowrates package-type equipment is available.

A good example of an application where the settler is an integral part of the process system is interstage clarification of phosphoric acid in the wet process. Since the clarification takes place between two evaporation steps, heat conservation is critical. With an inclined plate settler, there is almost no heat loss. Materials of construction for this severe duty are stainless steel or rubberlined carbon steel tanks and plates of stainless steel or special grade FRP. The exotic materials make an inclined plate settler very attractive compared to conventional settlers. A recent move — shared by many other CPI processes — is to cover the settlers and withdraw the fumes to scrubbers, accentuating more the advantages of compactness (6).

Other process — type applications include clarification of chloralkali brines and clarification/thickening of other brines. Still another is clarification of caprolactam under a nitrogen blanket (7).

With the introduction of the compact settlers the settler could often be located close to a source of wastewater and valuable product be recovered from the wastewater before it was combined with other wastewaters. Several inclined plate settlers are installed to recover various types of pigments. Another example is recovery of catalyst fines.

There are numerous plate settlers included in various wastewater treatment schemes, *e.g.*, chromate removal from cooling tower blowdown, removal of carbon and graphite fines, separation of plastic fines, removal of tetraethyl lead, pumice water treatment, ceramics wastes and various metallic waste streams. For all these applications various pretreatment schemes are used — the pretreatment is generally the most critical part of the separation process — and the inclined plate settler is basically replacing a conventional clarifier or thickener most of the time working at comparable surface loading rates and producing comparable results.

Early misapplications of inclined plate settlers on biologically treated wastewaters have given plate settlers a bad name for such applications. There are, however, successful experiences in this field too. Depending on the type of biological process there may be a need for periodic backflushing of the plates (once every 2–4 weeks for activated sludge). This can be done without shutting down the whole plant; an automatic system can be used for large plants.

A NEW CONCEPT — THE LGST

During the last few years a new concept of inclined plate settling has been introduced. The LGST (Lamella Gravity Settler/Thickener) is basically a clarifier

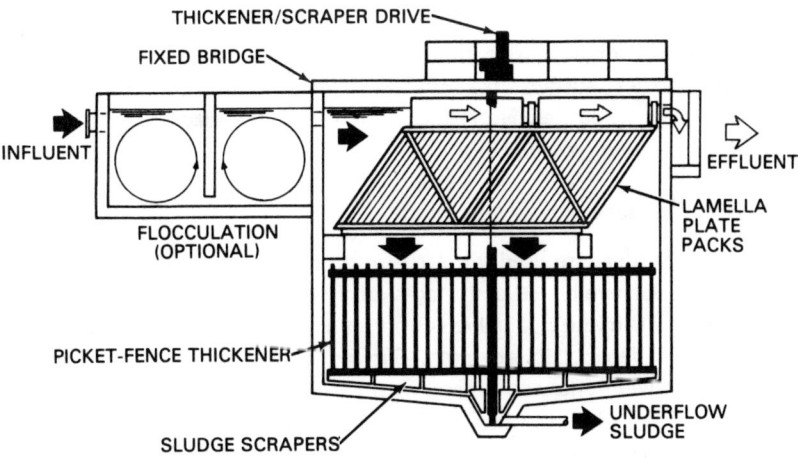

Figure 7.9 LGST. The plates are mounted in indepently supported and removable plate packs.

over a thickener. The top plate section consists of plates fed from the side to avoid the feed from entering the sludge compartment (Figure 7.9).

The LGST has a deep sludge compartment for compression containing a picket fence rake mechanism for sludge transport and to physically work on the sludge. It is obvious that this kind of sludge compartment is a positive feature for all applications for which compression is vital to achieve the desired underflow concentration. The LGST also offers considerably more storage capacity than hopper-type equipment.

It follows that the LGST offers advantages for several different types of applications. However, the most spectacular advantages are present for clarification/thickening of dilute metal hydroxide suspensions, such as surface water treatment using alum or ferric salts as coagulants. For such applications there have been recorded underflow sludge concentrations and properties that are not known to have been produced by any conventional thickeners. The characteristics of LGST thickening are as follows:

1. There is a perfect distribution of solids over the cross section of the compartment since the same amount of solids is being discharged from the bottom of each individual plate. Compare with a conventional thickener where all solids are introduced via the centerwell.
2. The solids are discharged from each plate into a completely stagnant sludge compartment. The solids are not disturbed by any turbulence. Compare with a conventional thickener where flow patterns are present over the whole cross section of the thickener compression zone.

Thus the sludge compartment of the LGST is an ideal thickener for which the whole depth is used for compression. In addition it turns out that metal hydroxide flocs are very susceptible to the kneading action of the moving picket fence mechanism in this ideal thickener. The flocs are destroyed and the work of the picket-fence mechanism will, with time, change the floc structure toward a more ordered form with less surface area, resulting in higher sludge concentrations and improved filtration properties. No reflocculation will take place in the sludge bed since stagnant conditions prevail (8).

Surface Water Treatment

Inclined plate settlers are used for clarification of surface water both for process and potable use. Generally, several stages of flocculation are used with a total detention time of 10 minutes if polyelectrolytes are used as coagulants, and 20–30 minutes if alum or ferric salts are used. Surface loading rates are comparable to what is used in conventional equipment, *i.e.,* 1.0–1.5 m/hr (0.4–0.7 gpm/ft^2) for alum or ferric and 2.0–2.5 m/hr (0.8–1.0 gpm/ft^2) for polyelectrolyte only.

Conventional thickening of low-turbidity alum sludges usually yields a sludge of 2.5–3.0% with a maximum of 5%. Ferric sludges generally yield somewhat higher concentrations. The LGST technique yields sludge concentrations of 10–20% and filtration rates at least three times as high as for conventionally thickened sludge. In addition the amount of sludge conditioner needed for dewatering can be cut dramatically. As the dewatering section of a surface water treatment plant often is more expensive than the total coagulation — clarification — thickening section, it is easy to imagine the advantages of the LGST concept.

REFERENCES

1. Forsell, B. and Hedstrom, B., "Lamella Sedimentation: A Compact Separation Technique," *J. Water Poll. Control Fed.* 47: 834 (1975).
2. Yao, K. M. "Design of High-Rate Settlers," *J. Environ. Eng. Div.* (October 1973).
3. Gebauer, A. G. *J. Environ. Eng. Div.* (October 1974).
4. *Process Design Techniques for Industrial Waste Treatment* (Enviro. Press, 1974), Chapter 8.
5. Jernqvist, A. "Lamella Thickening — An Attempt to a Theoretical Analysis," *Acta Polytech. Scand.* (1967).
6. York, R. "Phosphoric Acid Sludge Removed Efficiently with Inclined — Tray Settler," *Chem. Proc.* (May 1976).
7. Fischer, M. C. "The Role of the Lamella Gravity Settler in the CPI," paper prepared at AIChE Meeting, Philadelphia, PA, June, 1978.
8. Grebauer, A. G. and I. V. Janerus. "Gravity Thickening of Water-Treatment Plant Sudges, *J. Am. Water Works Assoc.* 70 (1):47 (1978).

CHAPTER 8

STRAINER SELECTION

ROBERT C. REISENWEBER
Zurn Industries, Inc.
Fluid Handling Division
Erie, PA.

Many process plants require large amounts of raw water from lakes or streams for cooling and component protection services. This water must be suitably treated to insure particles are removed which could jam or plus a component or cooling tube, causing loss of fluid and resulting in component failure or plant shutdown. Likewise, some process fluids require sizing or retention of solids which would be detrimental to the end product. In some cases, the solids retained are the end product.

Except for retention of very fine particles where a filter with disposable elements is used, most applications can be satisfed with a coarse straining device. This device, the pipeline strainer, is applied where solids in the range of 100 micron and larger are to be removed.

The purpose of this chapter is to describe the different types of pipeline strainers and develop the selection process used to properly size and apply them.

The simplest type of strainer is a plate with holes smaller than the particles to be retained, placed between two flanges in the pipeline. This type of strainer is the FLAT. A covering of wire mesh may also be used for finer straining applications, but even with coarse openings, the flow area of the pipe is reduced by at least fifty percent. As the flat presents a considerable obstruction to the fluid flow, its use is normally limited to start-up applications. When the new pipeline is flushed at start-up, the flat will stop large debris, such as pieces of weld rod or cigarette wrappers, from damaging pumps or other components in the line. This type of strainer is termed a "temporary strainer" and is normally removed from the pipeline after the initial flushing is completed.

Another type of temporary strainer is the CONE (Figure 8.1). This is a perforated plate or wire mesh that is supplied in a conical shape which provides more straining area lessening the flow restriction. It is mounted as shown in what is called a "spool piece." This is a length of pipe with flanged ends to house the Cone. As with the Flat, the Cone is removed from the line after start-up and the Cone with its retained debris is discarded. The spool piece is reinserted in the pipeline and the Cone strainer is put in storage until the line is shut down for major repairs or overhaul.

Occasionally, these "temporary" strainers are used in applications as "permanent" strainers. Permanent in the sense that they are left in the pipeline during

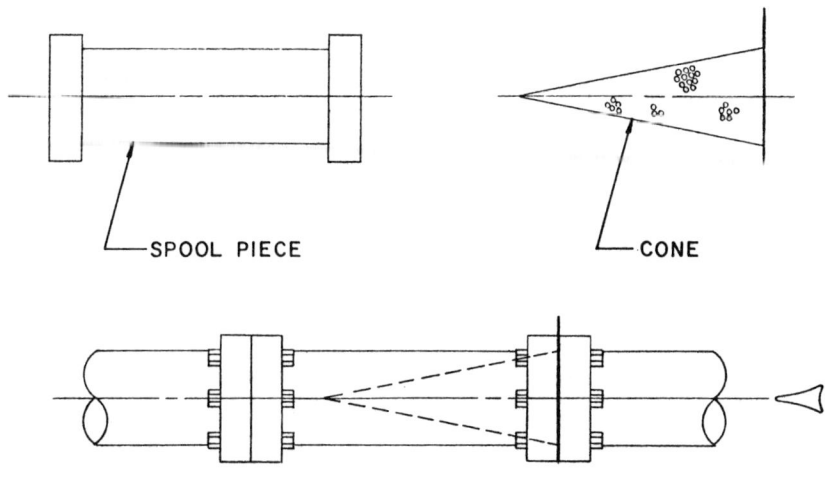

Figure 8.1

operational periods. This decision is usually justified based upon the initial low cost of these strainers. However, with rising maintenance costs and the ease in which permanent strainers can be serviced, the Flat and Cone Type Strainers soon lose this economic advantage.

A strainer which lends itself to both temporary and permanent applications is the Tee Type strainer (Figure 8.2). Fabricated out of pipe, the Tee is bolted or welded in the pipeline. Its straining element which rivals the Cone for area, is called a "Bathtub." The branch of the Tee has a cover which allows easy and quick access to the bathtub for servicing.

Similar to the Tee strainer is the Y-Type strainer (Figure 8.3). This type utilizes a cylindrical straining element called a "Sleeve." Sleeves are available in perforated plate or wire mesh and are accessible without removing the strainer from the pipeline. The Y-Type strainer can also be obtained in a wide variety of materials of construction, and is used in relatively clean systems to stop and retain the occasional piece of pipe scale which would foul the line.

The simplest of the permanent strainers is the single basket strainer (Figure 8.4). This strainer is referred to as a Sinlex Type strainer. Its straining element is in the shape of a basket which has a handle for removal of the element from the body cavity (Figure 8.5). The shape of the basket with its large screen area provides greater debris holding capacity. The larger the holding capacity, the longer the time between cleanings. This is the main advantage over the "Y" and Tee Type strainers. Another advantage is that the configuration of the strainer allows basket removal without complete draining of the pipeline fluid.

With all previous strainers discussed above, temporary or permanent, the pipeline

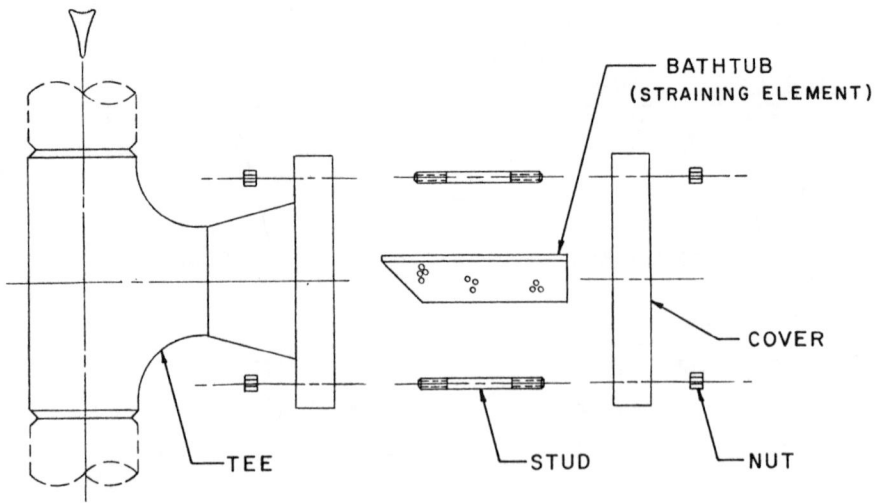

Figure 8.2

Figure 8.3

Process Equipment Series Volume 1

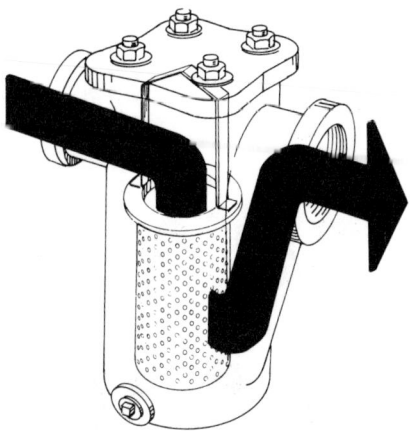

Figure 8.4

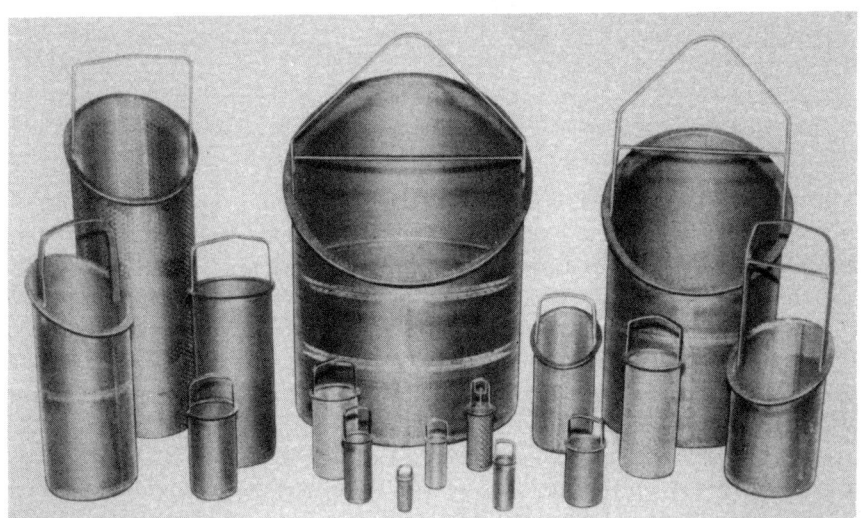

Figure 8.5

has to be shut down to clean the straining element. This is easily accomplished by isolating the strainer with inlet and outlet valves along with a means to de-pressurize the unit. But, this limits the application of these units to lines in which the flow can be interrupted for cleaning.

On pipelines where the fluid flow can not be interrupted, the Duplex Type strainer is used (Figure 8.6). Basically, the duplex is two (2) basket strainers linked with a flow diverting valve which allows access to the dirty basket while maintaining flow through the clean basket.

231

Chapter 8. Robert C. Reisenweber

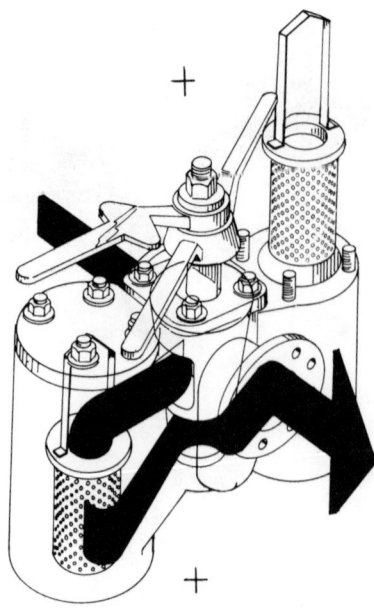

Figure 8.6

In all of these configurations, the straining element is cleaned manually. In continuous flow applications where the fluid may be extremely dirty or maintenance personnel may not be available, the Self-Cleaning strainer is used. This type of strainer has a number of elements which strain the fluid and collect debris. A rotating device, vented to atmosphere, is used to isolate the dirty elements. A portion of the strained fluid is backflushed through the isolated element, carrying the debris from the strainer. The pipeline flow is continuous and clean elements are constantly being exposed to the dirty fluid.

The self-cleaning strainer is considered expensive when comparing initial installations costs to the other types of strainers. However, self-cleaning strainers are normally used in critical pipelines handling large volumes of water which if interrupted, would cause extensive component damage or shutdown of the plant.

With these strainer configurations in mind, the first step in the selection process is the determination of the physical requirements of the application. These requirements include the degree of straining required, the pressure and temperature limits of the pipeline and materials of construction.

Unless the purpose of the strainer is to remove solids of a given size in a process fluid, the largest size of particle that can safely pass through the pipeline without causing damage or affecting line performance establishes the degree of straining. In this case, the strainer's sole function is to provide protection of components and care should be taken not to retain particles smaller than necessary. Retaining particles smaller than necessary results in increased element cleaning frequency which adds to operating and maintenance costs. The opening of the screen element should be approximately half the size of the opening to be protected. For example, a screen with 4 mm diameter perforations would be used to protect a spray nozzle which has a 8 mm diameter opening.

Also, the total area of screen holes, called free area, must be considered when determining proper sizing and selection of the type of strainer. This free area is compared to the pipe area and gives an indication of storage capacity. Typically, manufacturers build basket strainers so that coarse elements of 1 mm to 10 mm diameter openings will give a ratio of approximately six to one. The basket free area is six times the inlet pipe area and this provides the storage capacity necessary to provide reasonable time between servicing.

With fine straining applications where the screen opening is less than 1 mm in diameter, wire mesh supported by a coarse perforated plate is used. The resultant basket free area may be reduced as much as 50 percent, lowering its storage capacity. The open area ratio can be less than three to one on these fine straining applications and if the amount of anticipated solids in the fluid is low, the time between cleanings will still be acceptable.

After determining the size of the hole in the screen element, its effect on the open area ratio of the strainer configuration desired must be checked. Tee strainers and "Y" Type strainers may approach a two to one open area ratio in coarse applications, however, the ratio can become severely limited as finer straining is tried. If large amounts of solids in the fluid are anticipated, a successful application can be obtained by straining only as fine as necessary and using the strainer configuration which will provide the greatest storage capacity.

Series straining can also be used to obtain acceptable performance. Remove the large particles with a coarse strainer and follow it with another unit using fine elements to remove particles to the desired level.

A pipeline will be designed to a maximum pressure and temperature limit. These limits are imposed on all components and normally exceed the actual pressure and temperature which the system will be operated at.

Manufacturers of some components will rate their products as conforming to the pressure and temperature limits of their connections. As an example, a 15 mm (6") Class 150 flanged fitting can be operated at a pressure of 20 bar (290 psig) at $-28°$ C ($-20°$ F) or 1.3 bar (18.8 psig) at $540°$ C ($1000°$ F).

However, strainers are designed as a pressure vessel. The design and temperature limits are fixed in accordance with the closure style that allows access to the straining element. The pressure and temperature limits on the strainer nameplate should be checked. It must not be assumed the pipeline pressure can be increased by lowering the operating temperature or vice versa.

The materials of construction are normally the same as other components in the pipeline. Corrosion resistance requirements, severe shock and vibration conditions or very high pressures and temperatures can affect and limit the materials selection. As long as the strainer housing can be rated to the pressure and temperature requirements of the application, the material selected can be the same as the associated valves and pumps.

The last step in the selection process is the evaluation of the resistance to flow or pressure drop of the strainer as compared to the maximum allowable pressure drop of the system.

Manufacturers generally design strainers to match pipeline sizes. For example, a 15 cm. (6") strainer will fit a 15 cm. (6") pipeline. However, in some instances, the pressure drop of the strainer, depending upon the fluid flow, can exceed the maximum allowed by the system design. For example, if the device to be protected by the strainer requires a fluid pressure of 4 bar (58 psig) to operate, and the pressure available is 6 bar (87 psig), then the pressure loss caused by inserting the strainer must not exceed 2 bar (29 psig). This 2 bar (29 psig) is the maximum allowable pressure drop of the system.

When selecting the strainer, the manufacturers published pressure drop chart for the configuration of strainer desired should be checked to insure that the pressure loss does not cause the system loss to exceed the maximum allowed. Also, the selection should provide for the increase in drop as debris accumulates.

A typical pressure drop chart is shown in Figure 8.7. By knowing the maximum flow through the pipeline where the strainer is to be installed, the clean element pressure drop can be determined from the chart for the strainer size selected. As an example, assume a sinlex strainer is to be used with a flow rate of 570 m^3/h (2500 gpm) water. The dotted line in Figure 8.7 shows that this fluid flow of 570 m^3/h (2500 gpm) produces the following pressure drops:

> 20 cm (8") size strainer = .17 kg/cm^2 (2.4 psig)
> 25 cm (10") size strainer = .07 kg/cm^2 (1 psig)
> 30 cm (12") size strainer = .03 kg/cm^2 (.4 psig)

The size of the strainer selected from Figure 8.7 will depend on the maximum allowable pressure drop, the size of the straining media selected, and the solids content of the fluid being strained. Assuming the pressure drops shown do not exceed the maximum allowable pressure drop for the system, the 20 cm (8") size strainer can be used if the straining basket is coarse (3 mm diameter and larger) and the solids content is very low (fluid has been pre-strained and occasional pipe scale is to be removed). The 25 cm (10") strainer is applicable if the basket has coarse to medium size openings (as fine as 20 mesh wire) and the solids content medium (continuous amounts of debris during the straining period are expected). But if the straining is to be fine (40 mesh wire and smaller) or the solids content high (the fluid is not pre-strained and slugs of debris common), the 30 cm (12") size is a better choice.

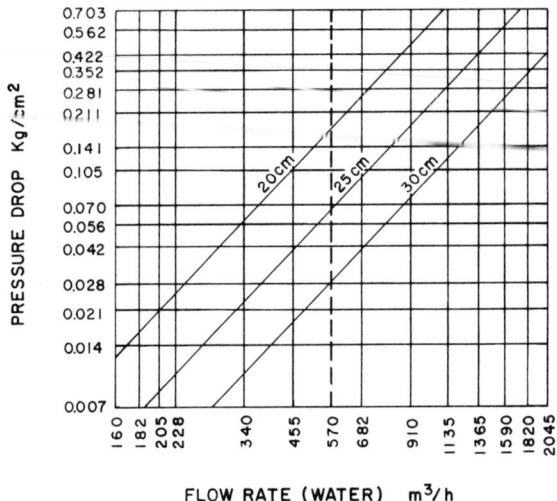

Figure 8.7 Pressure drop chart.

Manufacturer's pressure drop charts are based on the fluid being water. When more viscous fluids such as fuel oil are being encountered, contact the manufacturer for assistance in selecting the size of strainer.

While the pressure drop or pipeline size of self-cleaning strainers may indicate the size of strainer to be used, the velocity of fluid flow in the pipeline can become a determining factor. The manufacturer will evaluate this flow with regard to the strainer element free area to insure that flow through the element is not excessive.

This consideration prevents the debris being strained from being too tightly packed in the straining element and thereby allows quick and efficient removal of debris during the backwash or cleaning cycle. The backwash line from the self-cleaning strainer should be as short as possible, allowing a free-flowing discharge with a minimum of valves and elbows, to insure unit effectiveness.

In summary, selecting the right pipeline strainer for a given application is a step-by-step process of determining first, the physical requirements of the system, such as; the degree of straining required, the pressure and temperature limits, and the acceptable materials. To these requirements are added the desired performance considerations of the system which dictate the type and size of the strainer to be used. In general, there are no bad strainers, only bad applications. Following the process described in this chapter will help to insure the right strainer selection for the intended application.

CHAPTER 9

FLUID MIXING

J. Y. OLDSHUE
Mixing Equipment Co., Inc.
A Unit of General Signal Corp.
Rochester, NY

INTRODUCTION

Mixing encompasses a tremendous variety of operations. For the most meaningful discussion we should go one step further and ask "What type of mixing result are we trying to achieve?" Table 9.1 gives a definition of the kinds of mixing jobs that can be achieved. In the center of the table is the breakdown into the types of fluids being handled. Within each type of fluid system there are two very different mixing objectives:
1. physical uniformity, which includes such things as blending and solid-suspension, in which the criterion of success or failure is the uniformity achieved; and
2. mass transfer and chemical reaction processes, which involve the transfer of material from one phase to another, and include such things as gas absorption and liquid-liquid extraction.

Table 9.1. Mixing Processes

Physical Processing	Application Classes	Chemical Processing
Suspension	Liquid-solid	Dissolving
Dispersions	Liquid-gas	Absorption
Emulsions	Immiscible liquids	Extraction
Blending	Miscible liquids	Reactions
Pumping	Fluid motion	Heat Transfer

VARIABLE BATCH SIZES

Before proceeding to a discussion of the basic principles and practice of fluid mixing, it would be well to treat one of the most important aspects of obtaining satisfactory fluid mixing in an actual installation. If there are a variety of batch sizes and/or various batch viscosities to be handled in a single mixing vessel, the requirements must be specified carefully. Normally there is only one mixer drive and mixer shaft. There may be one, two or more impellers, and there may, or may not, be variable speed.

At a constant mixer speed, the power drawn by a single impeller is approximately constant once it is adequately covered. If there are two or three distinct batch levels to be used, then it is often possible to use two or more impellers so that each impeller is properly covered for a particular batch depth. By varying the diameter, blade dimensions or impeller type of each impeller, it is possible to get widely different degrees of mixing at constant speed at two or more distinct batch levels. It is very difficult to eliminate splashing and spraying when the final or transient liquid level is right at the impeller level.

If the requirements of the different batch levels are markedly different, it may be desirable to have two-speed drives, four-speed, or variable speed. It is also possible to install different numbers and kinds of baffles at various batch levels to take care of particular requirements.

In essence, there is hardly any combination of batch levels, viscosities and mixing requirements that cannot be handled if they are completely specified when initially designing the mixer. All the principles discussed in this chapter apply to any particular batch level and the impellers that are fully operable at that condition.

MIXING PROCESS DESIGN

From a practical standpoint, there has to be an accurate description of what mixing is required before a mixer selection can be made. Quite often this description is based on presently operating equipment, either a presently satisfactory operation or the desire to make a change or improvement. It would not be practical to try and list absolute conditions for the tremendous variety of mixing requirements encountered. The approach taken here is predicated on the assumption that there is a knowledge about a particular operation being carried out, or that there is a similar operation available, and the question is, "What mixer is required for a different batch size, incorporating any desired changes or improvements?"

Types of Mixers

The first classification that can be made is between top-entering and side-entering. Side-entering mixers use propellers and operate at such typical speeds at 380, 420 and/or 1150 rpm. Side-entering mixers are particularly effective for blending processes in which there are no appreciable amounts of free-settling solids present. The flow pattern is shown in Figure 9.1. They are usually the most economical choice for relatively large-volume blending. Their main operating drawback is in the submerged seal or stuffing box where the shaft enters the tank. Extremely low-liquid-level operation is difficult.

Top-entering equipment is usually propeller type in the smaller sizes, up to three horsepower, and turbine type above three horse-power. Top-entering propeller units are often mounted angularly off-center, Figure 9.2, to give effective top-to--bottom turnover. For general blending and suspension operations, the axial flow turbine, Figure 9.3, is normally used.

Flow Patterns

Referring to Table 9.2, a distinction is made between fluid mechanics of flow

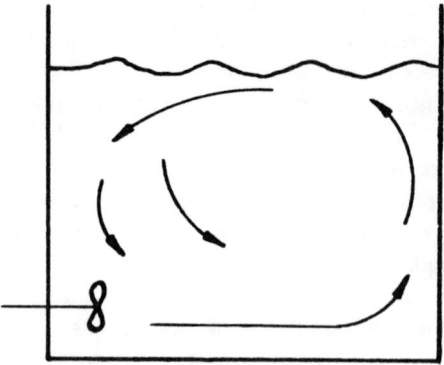

Figure 9.1 Typical flow pattern of side-entering, propeller-type mixer.

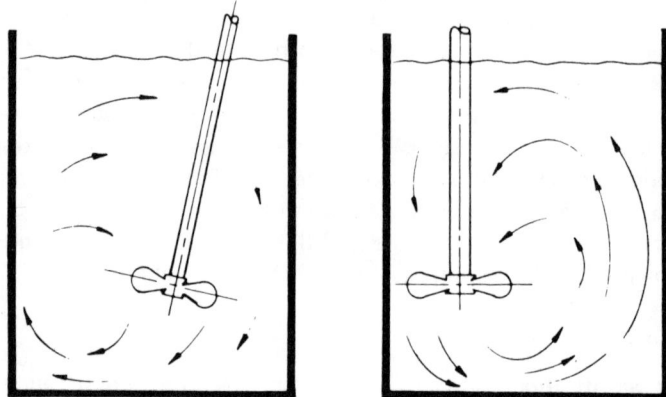

Figure 9.2 Typical flow pattern from top-entering, propeller-type mixer in off-center position.

from mixing impellers and the requirements of a given process for a particular flow pattern. There is no way to judge whether a given flow pattern is good or bad in the abstract. There must be a specific process requirement to evaluate the suitability of a particular flow pattern.

Measuring or predicting the flow patterns from various kinds of impellers is quite important in understanding mixing results. For example, the flat blade turbine can be made with or without a disc. It is sometimes assumed that the flow pattern for a turbine without a disc is radial out to the tank wall, as shown in Figure 9.4. In waterlike materials it seldom does act this way. Placed close to the bottom of the vessel, the impeller pumps downward like the axial flow turbine in Figure 9.5.

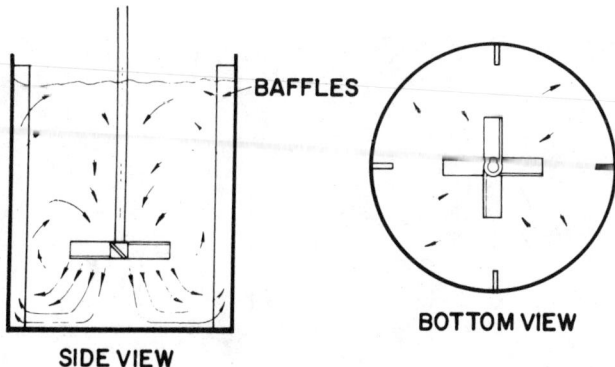

Figure 9.3 Typical flow pattern from axial flow turbine in baffled tank.

Table 9.2. Elements of Mixing Design

Process Design	1. Fluid mechanics of impellers
	2. Fluid regime required by process
	3. Scale-up; hydraulic similarity
Impeller Power Characteristics	4. Relate impeller, HP, speed and diameter
Mechanical Design	5. Impellers
	6. Shafts
	7. Drive assembly

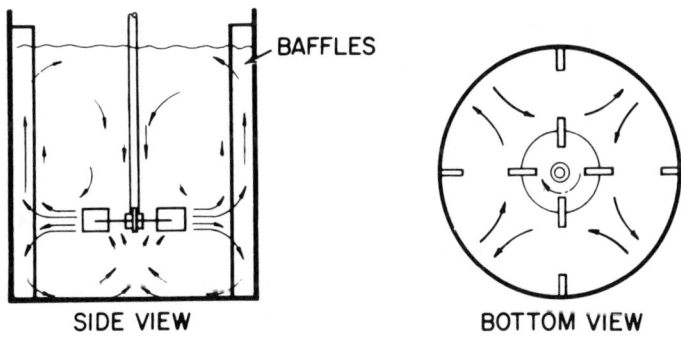

Figure 9.4 Usual flow pattern assumed for radial flow turbines.

Process Equipment Series Volume 1

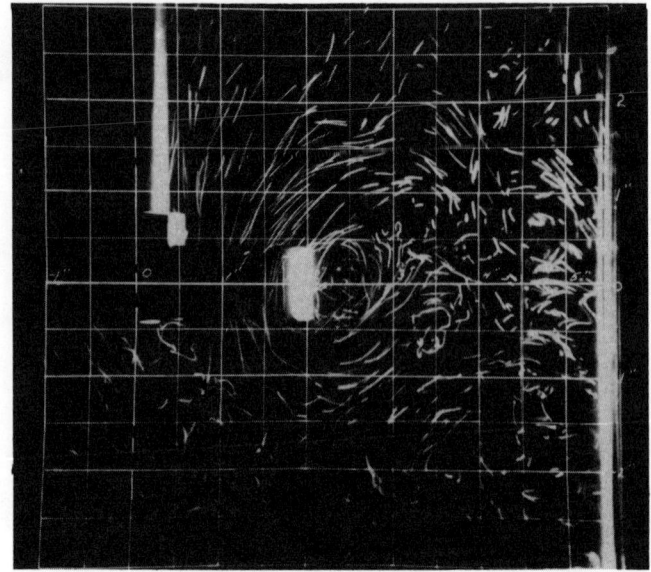

Figure 9.5 Streak photograph showing axial pumping characteristics for open flat-blade turbine.

Placed farther off the tank bottom, it pumps upward like an upward-thrusting axial flow turbine. This can make the prediction of process results quite difficult if a radial flow pattern has been assumed.

Baffles

Tanks may be classified as baffled or unbaffled in terms of their general overall fluid flow characteristics. A baffled tank has the following characteristics:

1. top-to-bottom turnover with complete intermixing of all fluid streams throughout the entire vessel;
2. the absence of a vortex through which air is drawn down into the fluid; and
3. the absence of a swirling flow pattern as contrasted to top-to-bottom turnover.

In most mixing cases a baffled flow pattern is desirable. There are times, however, when it is desirable to draw down powders from the surface or introduce gas from the surface; then an unbaffled flow pattern, Figure 9.6, is useful.

Baffled flow patterns at high power levels can often be unstable and cause severe fluid forces, which act on the impeller and the shaft and are erratic in scale-up performance.

A baffled flow pattern may be produced in several different ways. In a vertical, cylindrical tank the most common way is the use of four baffles, each one-twelfth

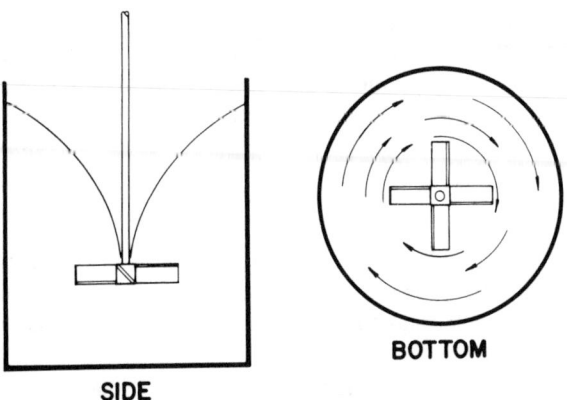

Figure 9.6 Typical unbaffled flow pattern.

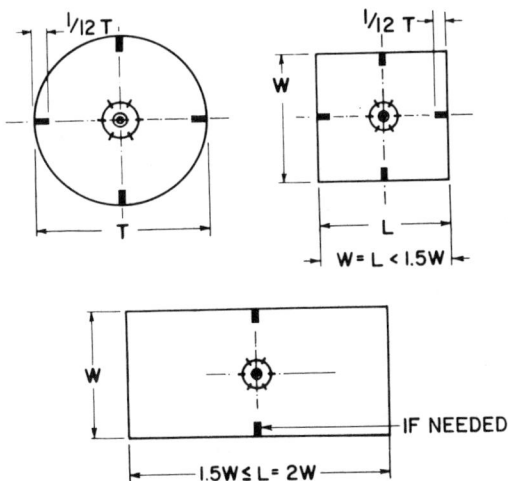

Figure 9.7 Schematic illustration of tank baffles.

the tank diameter in width, placed equally around the tank (Figure 9.7). This baffle width may often be modified when dealing with non-Newtonian fluids. With top-entering propeller type mixers up to about three horsepower capacity, the mixer can be placed in an angular off-center position and achieve the characteristics of a baffled flow pattern.

In a rectangular tank the shape can supply a baffling action at low power levels. A baffled flow pattern in a square tank is produced with two baffles placed opposite the impeller on opposite sides, each baffle being one-twelfth the tank width. This system is also used in tanks in which the length is less than one-half times the width. The corners act as partial baffles and, in some cases, baffles can be eliminated and still retain the top-to-bottom flow pattern. In rectangular tanks, in which the length is one-and-one-half times the width, it is possible to eliminate the use of baffles. At high horsepower levels baffles must be provided even in rectangular tanks. These are illustrated in Figure 9.7.

Impellers

Impellers may be divided into two general classifications—axial and radial flow. Typical flow patterns are shown in Figure 9.3. Many modifications can be made to the blade shapes of paddles and turbines, but their overall flow pattern in a baffled tank is basically very similar.

Considering the impellers, mention should be made that it is essential that an impeller produce the desired process result under conditions that allow sound mechanical design. Many of the proportions and designs that are used commercially were selected to achieve the desired process result under conditions of sound mechanical operation.

Flow and Shear Rate

The pumping capacity, Q, of propeller and turbine impellers for any given geometric series of impellers is given by the proportionality:

$$Q \propto ND^3$$

where N = impeller speed, rpm
 D = impeller diameter
 Q = volumetric fluid displacement of impeller

The power drawn by the impeller produces this circulating capacity against the impeller head across the system. This impeller head is related conceptually to the fluid shear rate existing in the tank.

At a constant power level the pumping capacity of various diameter impellers of the same geometric proportions in a given tank is related by the proportionality:

$$Q \propto D^{4/3} \text{ (Constant Power)}$$

However, the flow from the impeller entrains fluid in the tank so that the total flow circulating through the system can be several times greater. As the impeller gets larger and larger, it is pumping more flow at less head. It also has less entrainment distance due to its proximity to the tank wall. The total flow in the system does not increase as fast as the above equation, and at about 0.6 D/T ratio, there is no further advantage gained in increasing the impeller diameter (Figure 9.8).

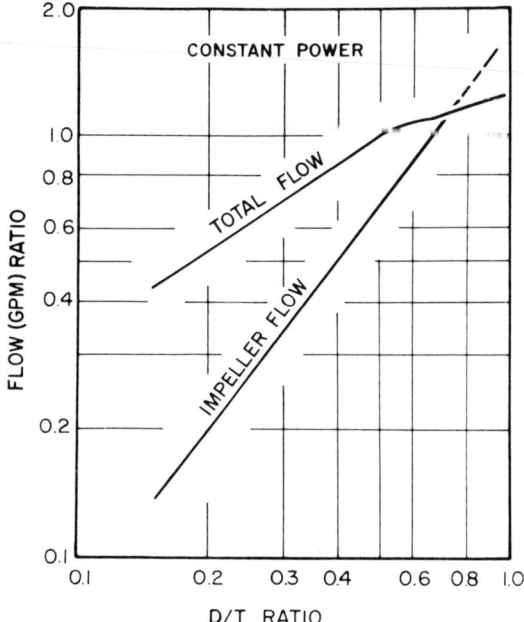

Figure 9.8 Impeller flow and total flow as a function of D/T ratio at constant power.

For high-viscosity applications it is often necessary to go to close-clearance impellers, which have a D/T ratio of 0.95 or greater, in order to positively pump fluid through the vessel and off of the tank walls.

The discussion so far holds most exactly for low-viscosity fluids. As viscosity increases, and/or if there are appreciable pseudoplastic effects present, there are certain minimum D/T ratios that must be used to get motion throughout the tank. Within the range of D/T's that can be used, most blending and suspension processes require less power with large-diameter impellers. Figure 9.9 illustrates typical performance. Even though power decreases with large impellers, the low speed of these impellers actually requires an increase in torque for the mixer drive. A balance between first cost and operating cost of the equipment is required.

Turning now to fluid shear rates, a mixing vessel has a complete spectrum of shear rates. Large-sized eddies and swirls transfer their energy to smaller eddies and down to still smaller-sized eddies, and eventually through viscous shear into heat. The essential requirement when examining the fluid shear required for a particular mixing process is to identify the size and frequency of the shear rates that affect that particular process result. We can illustrate the point by mentioning that we do not use an ultrasonic drill to drive a rivet, nor do we use a rivet hammer to drill a tooth.

The velocity profile from a radial flow turbine impeller is illustrated in Figure

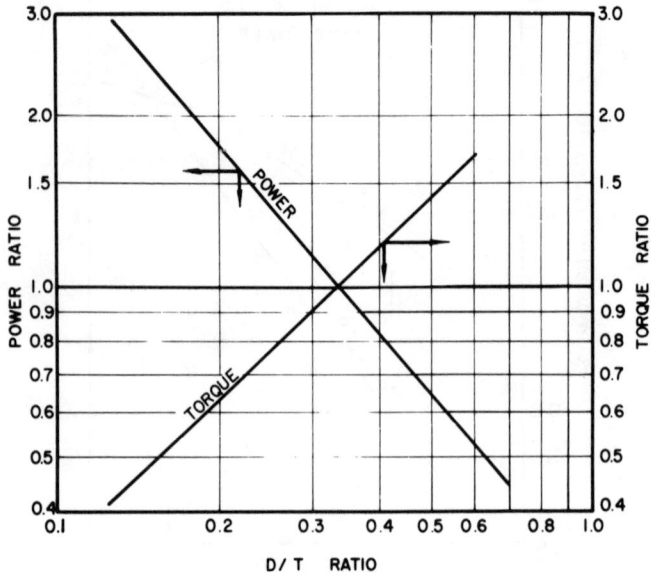

Figure 9.9 Effect of D/T ratio on typical blending process.

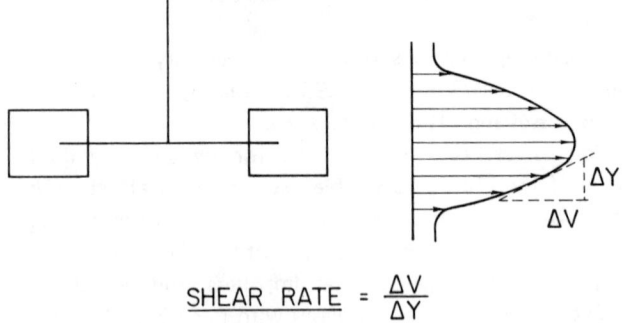

$$\text{SHEAR RATE} = \frac{\Delta V}{\Delta Y}$$

Figure 9.10 Schematic drawing of velocity profile from turbine impeller.

9.10. The slope of a line tangent to the velocity profile gives the shear rate at that point. Multiplying the shear rate by the viscosity of fluid at that point gives the shear stress. This shear stress breaks up particles and makes dispersions. Some examples of some of the recent research in this area are given later in this chapter.

If a particular fluid has a constant proportionality between the fluid shear stress in the fluid and the shear rate generated in the fluid, it is a Newtonian fluid and consideration of its viscosity is quite straightforward.

If, however, the viscosity of the fluid varies with shear rate, we then must know what viscosity does the impeller "see" at shear rates existing around the impeller, and what viscosity does the process "see," which best describes the viscosity at the shear rates in the remainder of the tank, which affect the overall process result.

Pseudoplastic fluids can be relatively Newtonian at very high shear rates and at very low shear rates, while being very non-Newtonian at shear rates existing in a mixing tank or in a commercial viscosimeter. There is no limit to the complexity or the number of arbitrary constants in a mathematical equation used to describe this condition. However, for shear rates between 1 and 30 seconds^{-1}, which are the order of magnitude of shear rates in a mixing tank, the viscosity usually plots as a straight line on a log-log plot with shear rate, and that the "Power Law" relationship with one arbitrary constant, n, is adequate.

Shear Stress = K (Shear Rate)^n at a given shear rate,

$$\mu = \frac{\text{Stress}}{\text{Rate}} = K' \text{(Shear Rate)}^{n-1}$$

GENERAL SURVEY OF FLUID MIXING EQUIPMENT

Portable and Fixed Units up to 5 Horsepower (Figure 9.11A and 9.11B)

There is a general class of equipment, normally using propellers, which runs at either direct drive speeds of 1,150 or 1,750 rpm, or at single reduction gear drive speeds between 300 and 420 rpm. They may either be clamped on the rim of open tanks or mounted with a fixed assembly for either open- or closed-tank operation.

These units are the most economical, are usually used in tanks without baffles, and are rugged and long-lasting when applied within their process capability.

Side-Entering Mixers (Figure 9.11C)

These mixers are largely used for blending purposes. The side-entering propeller type mixer is very economical and established a very effective flow pattern in tanks of almost any size. Because the shaft seal is below the liquid level, its use in fluids without corrosive and erosive properties is usually ideal. These units are used up to 5 million gallons in water-like fluids.

Heavy Duty Top-Entering Equipment (Figure 9.11D)

From about 3 hp on up, heavy duty, top-entering units are used for the larger and higher power per unit volume applications. These units are normally fixed to a rigid structure or tank mounting. Either radial flow turbines or axial flow turbines may be used, depending on the application.

Speeds for the more common types of impellers vary in the 50–100 rpm range,

Chapter 9. J. Y. Oldshue

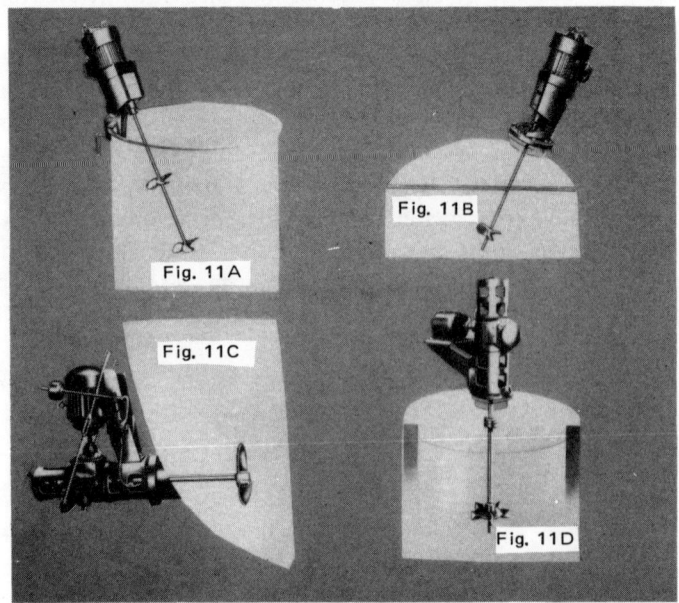

Figure 9.11 Photograph of diffent types of mixing equipment.

with speeds possibly going down to 30 rpm and up as high as 200 rpm. These usually require either a double set of helical gearing or a single set of worm gearing to achieve these speeds, so they are inherently more expensive than the single reduction units mentioned previously. Shaft sealing devices can also become quite elaborate, including stuffing boxes and high-pressure mechanical seals.

Slow-Speed, Close-Clearance Impellers

For use in "high-viscosity materials," helical (Figure 9.12) and anchor (Figure 9.13) close-clearance impellers are used at speeds from 5–20 rpm. Table 9.3 illustrates the comparison in power required and initial cost in the case that either the conventional open turbine or close-clearance impeller can accomplish a given job. Table 9.4 illustrates the viscosity ranges in which open- and close-clearance impellers are normally applied.

Lineblenders (Figure 9.14)

For retention times on the order of several seconds, mixers can be placed in a process pipeline. Usually the level of agitation in the lineblender is sufficient to completely disrupt the flow pattern going through it and make the fluid going through the unit experience one or two stages of what is commonly referred to as "perfect mixing." For use on gravity-flow systems, low-pressure drop is essential.

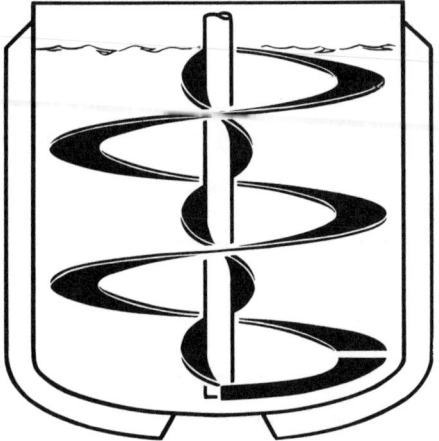

Figure 9.12 Schematic view of helical impeller in jacketed tank.

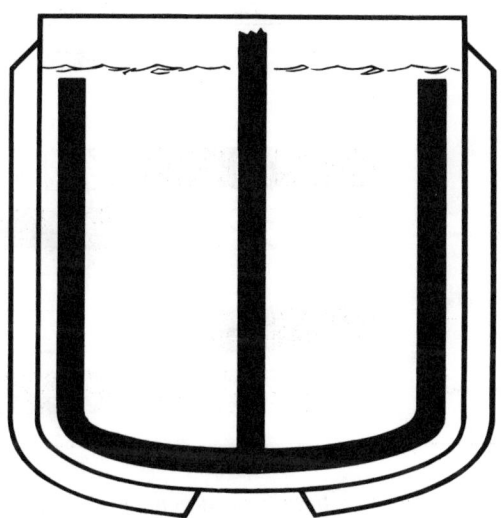

Figure 9.13 Schematic view of anchor impeller in jacketed tank.

Pressure drop specifications are important in this type of equipment and must be considered when selecting pumps for the process pipeline.

Aerators (Figure 9.15)

In addition to fully submerged mixing impellers with separate gas introduction, a

Table 9.3. Axial Flow Turbine, D/T=0.5, Compared to Helical Impeller ½-in. Radial Clearance

Impeller Type	Horsepower (Op. Cost & Mix. Heat Added)	$\dfrac{N}{\text{(RPM)}}$	Torque	Initial Cost
Axial Flow Turbine	1	1	1	1
Helical Impeller	1/4	1/8	2	3

Table 9.4. Typical Impeller Applications

large number of surface aerators have been developed. These usually can be classified as one of two different types. One type uses a relatively high-speed propeller unit, pumping liquid up and spraying it out through the air as compared to slower-speed, larger-diameter turbine impellers, which operate right at the interface causing spraying and entraining action right at the site. Both types can be either fixed or float-mounted in the aeration basin.

FLUID PROPERTIES FOR PROCESS SPECIFICATIONS

1. It is obvious that specific gravity of all components and of the mixture must be available.

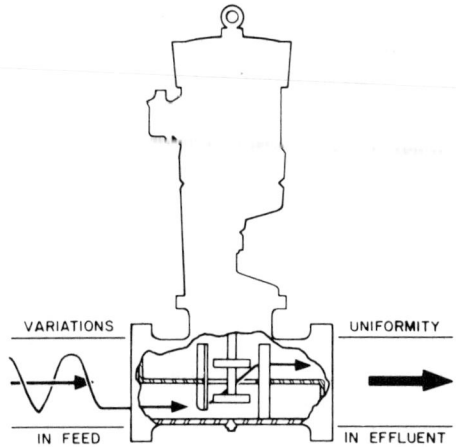

Figure 9.14 Schematic of line blender.

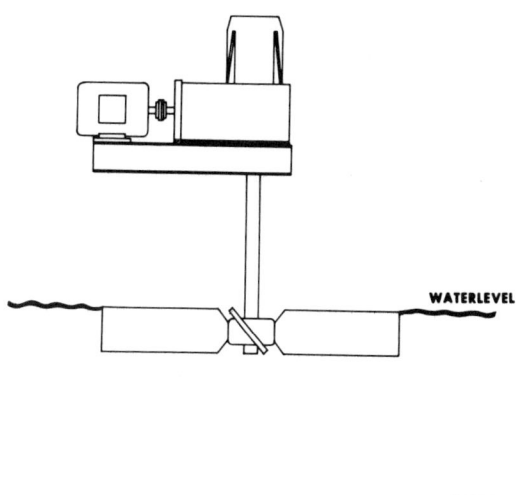

Figure 9.15 Schematic view of surface aerator.

2. The viscosity of the fluid is also important and the following information should be available.
   a. If the fluid is Newtonian, which means that it has a constant viscosity at all impeller speeds, then an approximate viscosity, either measured or estimated, is often satisfactory. Up to 5,000 centipoises estimated viscosities are often satisfactory. Above 5,000 centipoises, viscosity starts to play an increasingly

important role in the power consumption of the agitator, and estimating errors of 20—50% in viscosity can mean undesirable undersizing or oversizing.
   b. Temperature has a marked effect on viscosity and viscosity must always be given at expected operating temperatures in the plant.
   c. For non-Newtonian fluids, viscosity data are extremely important. Every geometric series of impellers has an average fluid shear rate related to its operating speed. For example, for a flat blade turbine impeller, the average impeller zone fluid shear rate is 11 times the operating speed. The economic importance of knowing the viscosity at the shear rate in the plant size unit governs how much time and effort is worth expending in obtaining the Power law exponent. In any event, measurement or extrapolation to actual operating plant shear rates must be made. If viscosities are measured by slow shear rate rotational viscosimeters, then there is always some error inherent in projecting these to higher shear rates in full-scale mixing equipment. The most exact way is to get the viscosity by using a standard mixing tank and impeller as a viscosimeter and, by measuring the power response on a small-scale mixer, obtain viscosity at shear rates similar to what they will be in the full-scale unit. Obtaining the exponent "n" in the Power law relationship allows calculation of viscosities at other shear rates.
3. Other physical properties are extremely important. Which phase is to be dispersed must be known.
4. For solid-liquid systems, settling velocities of the 10%, 50% and 90% by weight fractions of particle size distribution must be available. This will either have to be available from calculation or measurement. What is to happen with oversize solids will have to be considered. Different statements of suspension requirements will be covered later.

Gas-Liquid System

With gases, flowrate must be available at standard temperature and pressure as well as actual temperature and pressure. The range of gas flow must be given, and whether the mixer is to be operated at full horsepower for all the gas ranges or whether it is to be operated with the gas on. Additional comments on fluid properties will be given in each of the succeeding sections.

IMPELLER POWER CONSUMPTION

Before discussing process specifications, Table 9.2 illustrates the three basic areas of fluid mixer design. It is important at this point to distinguish between impeller power correlations, which hold true regardless of whether the process is being accomplished or not, and selection of variables from a process standpoint.

For example, for a given type of impeller and a given viscosity and gravity of the fluid, the power is related to the impeller speed and diameter. Thus, there are only two independent choices and only two variables are necessary for process correlations. For example, a heat transfer process correlation, shown in Figure 9.16, uses only speed and diameter. The power consumption must be calculated from the Reynolds number Power number curve shown in Figure 9.17.

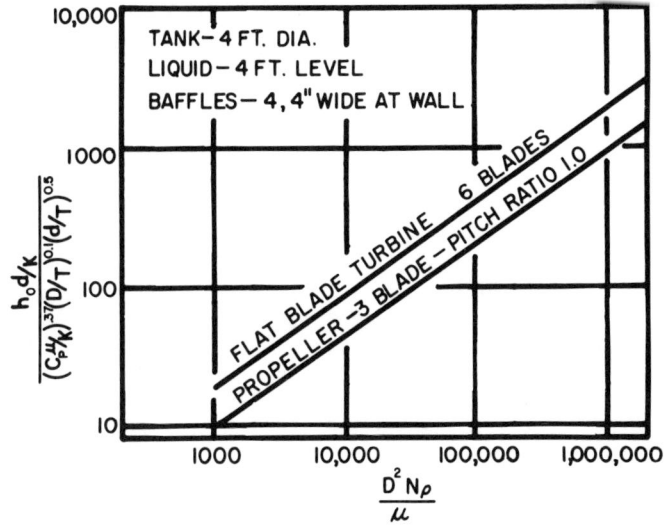

Figure 9.16 Heat transfer correlation for flat-blade turbine and propeller.

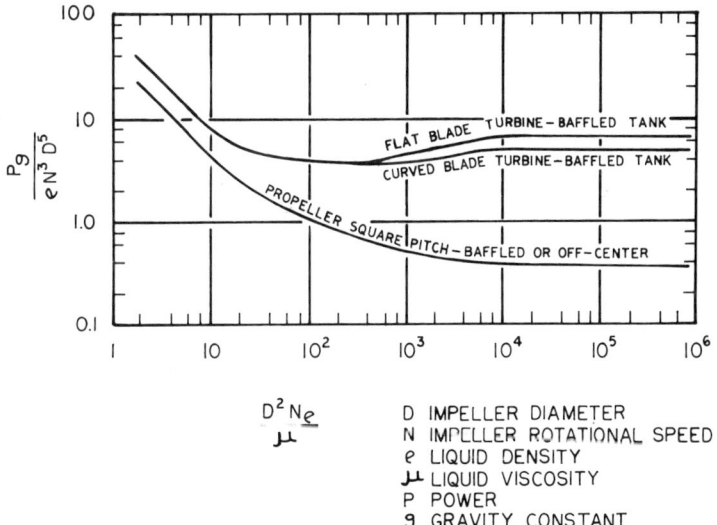

Figure 9.17 Reynolds number power number curve for flat-blade turbine and propellers.

Chapter 9. J. Y. Oldshue

PROCESS CRITERIA

The process discussion will be centered around the outline in Table 9.1. The first decision is whether the criterion can be expressed as a physical measurement of uniformity or whether the criterion involves a diffusion of material, either in mass transfer across a two-phase interface or a chemical reaction.

A second classification involves the five application classes shown in Table 9.1. There are 10 separate mixing areas listed. Examples of process specification in each of these 10 areas will be given.

Type of Data Needed

There are three possible situations that exist in a mixing application:

1. The process has been done before, and design is based on previous successful experience and data. It is then primarily a matter of getting the right fluid physical properties and defining the tank volume and shape. Reference to existing process installations is usually helpful in getting complete agreement on process requirements.
2. In some cases, scaleup data are available for the type of operation, but there must be an experimental run to obtain the actual process result obtained under a given set of conditions. For example, in a gas-liquid process involving a new process, reaction rate and mass transfer rate must be obtained at one or two conditions, but scaleup and design can be made from existing correlations.
3. A completely new process, differing in fluids and/or principles from any existing process installation. Several runs must be obtained on one tank size as well as properly designed runs on a second or third tank size to establish the scaleup relationship involved.

LIQUID-LIQUID MIXING

Liquid-Dispersion

There are many processes where liquid-liquid dispersion is the process criterion. For example, the size of a droplet in emulsion polymerization may be the most important parameter influencing the quality of the product. The role of fluid shear in these processes is critical. The ultimate possible particle size is determined by the tip speed of the impeller. If the system is established so that coalescence is prevented and stability is satisfactory in sufficient time, this will be the ultimate particle size that will result.

However, if the dispersion being produced is being used as an intermediary for carrying out a suitable liquid-liquid extraction, and the emulsion must be settled again, then a dynamic dispersion system is produced in which the maximum shear stress of the impeller determines what particle size can be produced in this zone while the average shear rate around the impeller and the average shear rate in the rest of the tank determines what the overall average particle size will be in a dynamic system.

Liquid-Liquid Mass Transfer

In these processes the liquid-liquid emulsion produced is primarily to provide interfacial area for mass transfer to take place plus possibly liquid-solid mass transfer steps. A choice of batch or continuous, single stage or multistage, and concurrent or countercurrent, illustrates the many different possibilities.

Mixing can affect the residence time distribution of both the continuous and dispersed phase. Figure 9.18 illustrates a typical countercurrent multistage column. One of the main features of a mixer column for liquid-liquid extraction is that the problem of transverse nonuniformity and unexpected and peculiar flow distributions can be avoided.

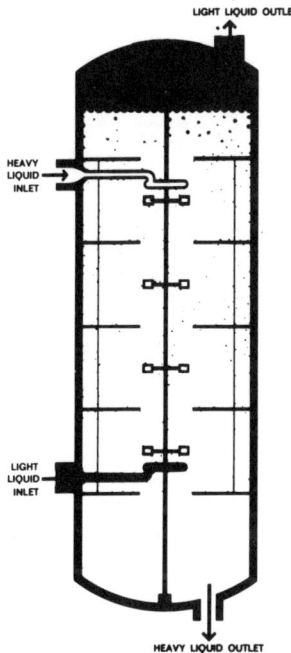

Figure 9.18 Schematic view of multistage extractor.

For liquid-liquid extraction there must be some data on the mass transfer rate of the system for typical mixing conditions. There must also be an applicable liquid-liquid equilibrium curve. If there is a chemical reaction occurring subsequent to mass transfer, data on this effect must be available.

MISIBLE LIQUIDS

Blending, Low Viscosity

Blending of similar low-viscosity materials can be a relatively easy operation.

However, if we are interested in analyzing in detail what blending is, or are interested in the degree of uniformity in blending high-viscosity materials, the "scale of scrutiny" of the uniformity of blending is critical.

Figure 9.19 shows the progress of blending on a batch basis from a tank which was initially stratified with hot and cold water. The impeller was in the lower cold phase as shown by Legend A. Legend B was about at the initial interface and C, D and E were in the hot layer. Since the probe used to detect blending shows uniformity at 105 seconds, this could be the blend time. If, however, we had a more sensitive measuring device or were interested down in the molecular scale, the blend time is considerably longer.

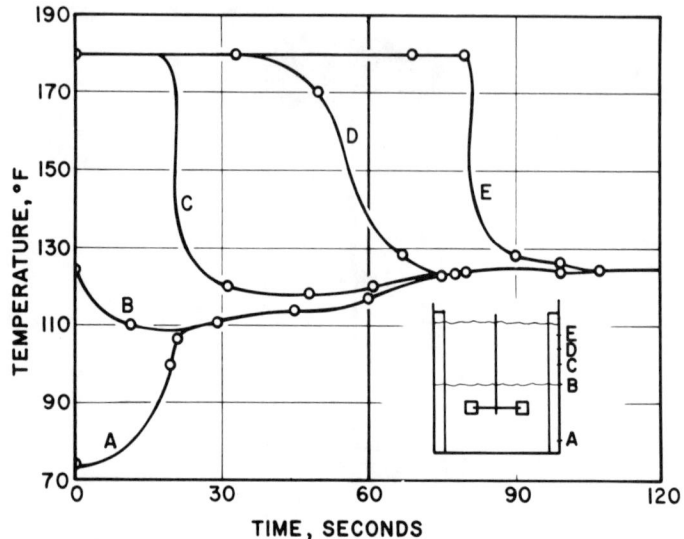

Figure 9.19 Progress of ble..ding in an initially stratified tank.

No matter what scale is used, there is always an absolute blending that is somewhat less than complete. Of course the ultimate definition of the blend or the molecular level depends on whether we use a random or ordered mathematical model.

Most blending operations follow the principles shown previously in Figure 9.9, in which less horsepower is required with larger-diameter impellers. Thus, there is always the relationship between operating cost and first cost to be considered.

For non-Newtonian fluids we first have to be sure that the entire batch is in motion. If this condition is satisfied, then blending progresses in the same fashion that it does in Newtonian fluid, and it has been found that blend time correlates well with the viscosity at the average shear rate of the tank with blend times obtained in Newtonian fluids.

Assuming that the blend time for a particular product is known, mixer power and blend time will be approximately inversely proportional (7). If the difference in specific gravities is changed, blend time goes up in proportion. Blend time is approximately directly proportional to viscosity changes in the continuous phase and to the ratio of viscosities between the phases.

If viscosity varies during a run, the mixer must actually be sized for the highest viscosity expected. At lower viscosities the mixer power level may be excessive, and splashing and excessive turbulence may be a problem. In cases where entrainment is a problem, a slower speed may be required at the low-viscosity stage.

In case it is desired to keep air entrainment completely out of a tank in which mixing is required, placing baffles across the surface of the bank, such that no splashing can occur over the top of the baffle, will eliminate this (Figure 9.20). These baffles should be high enough so that liquid will not wash over them and extend low enough so that the wave troughs do not extend below the baffle.

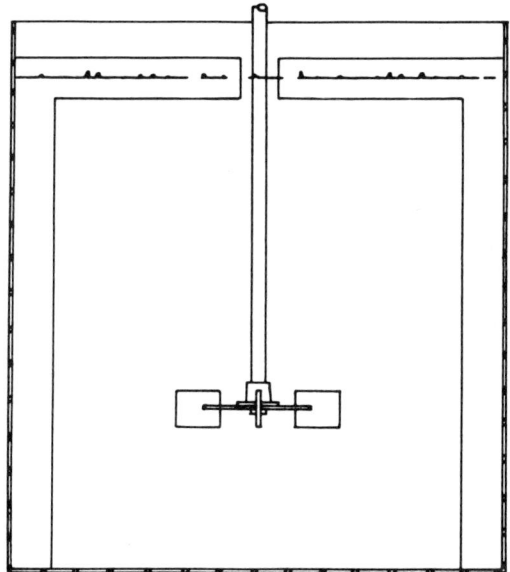

Figure 9.20 Horizontal surface baffles to eliminate air entrainment.

Heat Transfer

The transfer of heat across a tube in a low-viscosity fluid in a mixing tank depends entirely on the velocity of liquid around the tube. As a rule, the higher the pumping capacity from the impeller, the higher will be the heat transfer coefficient around the tube surface. Numerous articles (2–4), describe the effect of mixing on coils, jackets and vertical tubes.

Curves illustrating the difference between propellers and flat-blade turbines are shown in Figure 9.16. At equal Reynolds number in fully developed turbulent flow, there is a difference in the two impellers, but at equal power the performance of the two impellers is quite similar, as shown in Table 9.5.

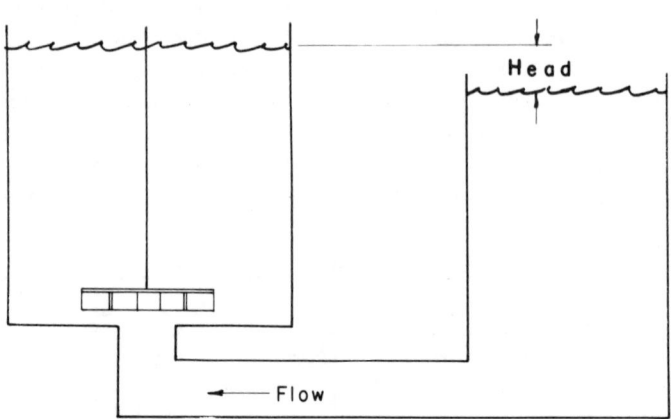

Figure 9.21. Typical installation, turbine, for pumping large volumes of fluid against low heads.

Table 9.5. Heat Transfer Coefficient for Propeller Compared with Turbine

$h_{o\ prop}/h_{o\ fbt}$ at constant N_{RE}	HP_{prop}/HP_{fbt} at constant N	$h_{o\ prop}/h_{o\ fbt}$ at same Diameter and constant N
0.54	0.06	1.0

$$N_{RE} > 10{,}000$$

Data on the thermal properties of the fluid are all that is necessary to calculate mixer side coefficients in most systems. For non-Newtonian fluids the heat transfer surface must be estimated and suitable viscosity obtained. As Reynolds number decreases, the transition area is approached. In going to close-clearance impellers, the mechanism of heat transfer turns from forced convection to conduction through a stagnant film. Measurements indicate that the heat transfer coefficient is given by the conduction through the film through a thickness of about one-half the radial clearance, so that for ½-in. radial clearance, in an organic fluid, a coefficient of about 4–5 Btu/hr/°F/ft² is obtained, regardless of impeller speed and viscosity.

Adding a scraper can usually double or triple a coefficient while also doubling or tripling the power consumption of the impeller.

FLUID MOTION

For pure pumping of fluids an impeller can be mounted above an orifice as shown in Figure 9.21, or in a side-arm circulator shown in Figure 9.22.

PROPELLER CIRCULATOR

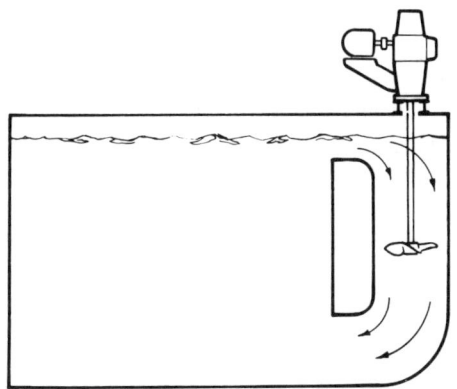

Figure 9.22 Typical installation, propeller, for pumping large volumes of fluids against low heads.

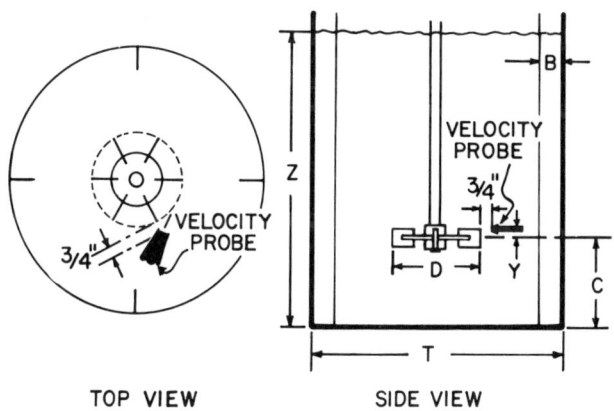

TOP VIEW   SIDE VIEW

Figure 9.23 Typical turbine installation in baffled tank, showing location of hot wire velocity probe.

It is necessary to determine the flow-head curves for the particular impeller as well as the flow-head curve for the system. Interaction of the system and the impeller characteristic determines the operating point for the application. Applications include keeping outdoor ponds free of ice, pumping water in flumes and channels and crystallizers.

SOME EXPERIMENTAL DATA ON FLUID SHEAR RATES IN MIXING TANKS

If the objective were to find only the mean velocity at a given point, then some modification of a pitot tube would be the most practical method. However, it was desired to determine the rapid fluctuations in a stream and so the hot-wire velocity meter was used as shown in Figure 9.23. The primary purpose here is to give the magnitude and ratios of some of the effects and to avoid becoming involved in complicated mathematical discussions of flow and turbulence. For simplicity, only the measurements made when the probe was positioned as shown in Figure 9.23 will be considered here.

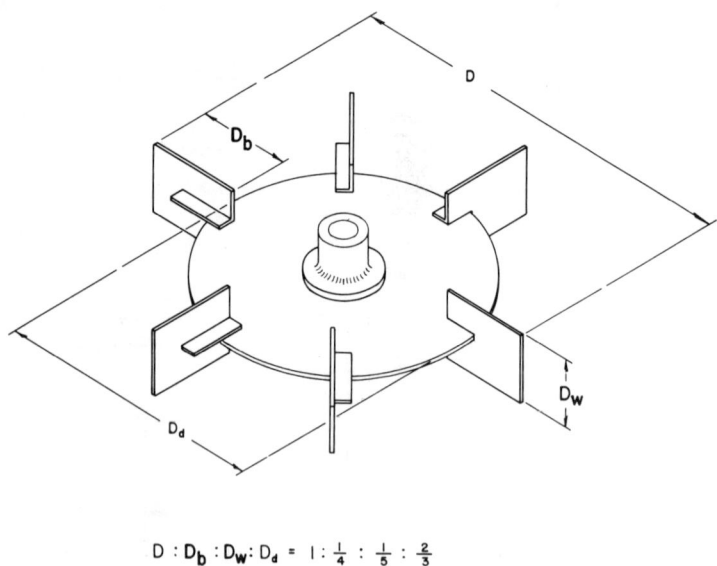

$$D : D_b : D_w : D_d = 1 : \frac{1}{4} : \frac{1}{5} : \frac{2}{3}$$

Figure 9.24 Typical dimensions of flat-blade turbine impeller.

The pumping capacity of flat-blade turbines has been measured by various investigators. The flat-blade disc turbine referred to here is of the type shown in Figure 9.24. It has six blades and the dimension ratios given on the figure.

One of the most complete works on pumping capacity is that of Sachs (5), who used the photographic technique of analysis for velocity patterns. To illustrate the type of flow pattern, Figures 9.25 and 9.26 are included from his work. They show the overall type of flow developed by a flat-blade turbine.

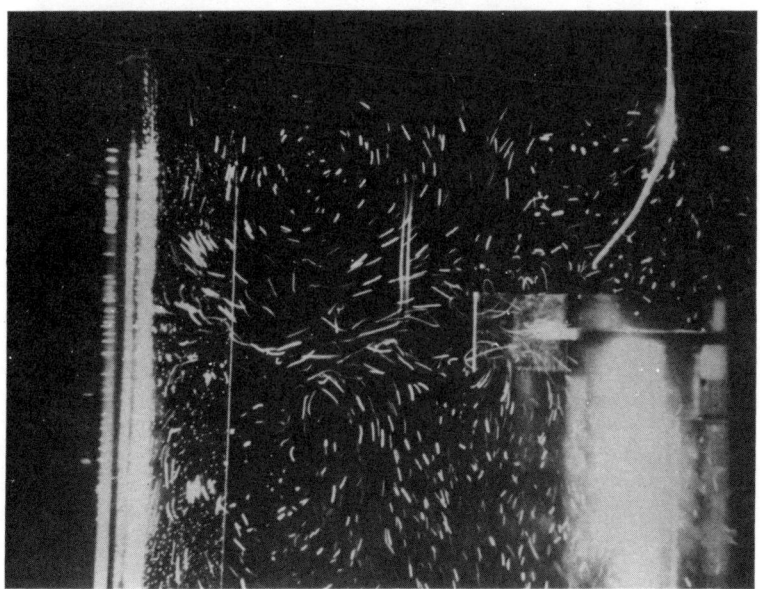

Figure 9.25 Flow pattern, side view, 4-in.-diameter flat-blade turbine: 12-in. diameter tank, water. Streak photography showing particle velocities in a ¼-in.- wide vertical plane.

To obtain more information on the nature of flow from a turbine impeller, a hot-wire velocity meter was developed for use in mixing systems. The measuring element consists of a small wire suspended between two electrodes. The wire was 0.100 in. long and 0.0007 in. in diameter. The wire was heated electrically to about 20°F above the tank temperature. As the water flows by the probe, it tends to cool it, and thus the amount of current required to maintain constant temperature is related to the velocity of the fluid.

The electronic arrangement of the equipment is as described by Hubbard (6). A hot-wire probe gives the maximum flowrates when the wire is at right angles to the flow stream. In this study, the wire was always positioned in a horizontal plane and was rotated at various angles to a radius from the centerline of the tank until maximum velocity readings were obtained. From this study, it appeared that the angle for maximum velocity corresponded to a tangential velocity component at the impeller periphery.

In the data reported here, the probe was positioned ¾ in. away from the turbine

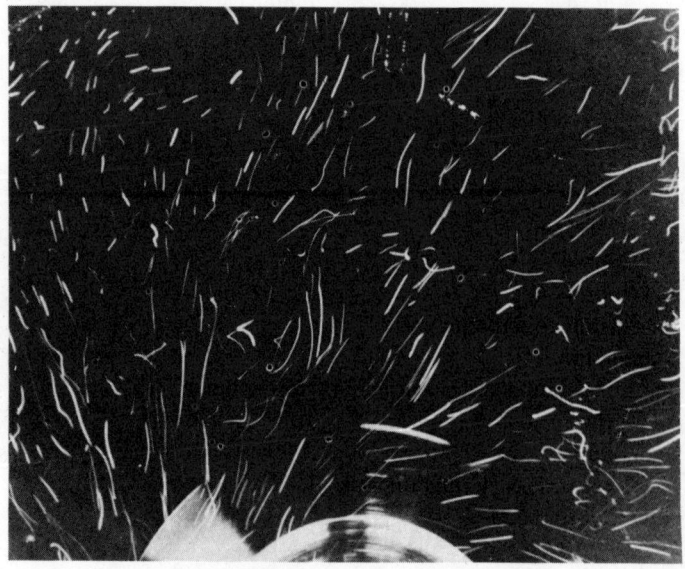

Figure 9.26 Typical flow pattern from flat-blade turbine in baffled tank. Bottom view. 4-in.-diameter turbine, 12-in.-diameter tank, water. ¼-in.-wide horizontal plane in impeller centerline.

periphery, directed along a tangent to the impeller periphery, and was positioned at varous distances above and below the turbine centerline.

The tank diameter, T, used in this work was 18 in. and the liquid level, Z, was 20 in. Impeller diameters, D, used were 4, 6 and 8 in. Four baffles were used in the tank, each 1.5 in. wide.

The hot-wire velocity meter has a very fast response, so that the velocity, u, at a point at any time can be expressed as a mean velocity over a time interval, $\bar{u}$, plus a fluctuating velocity component, u', so that:

$$u = \bar{u} + u'$$

The probe was calibrated by placing it into the flow from an orifice in a tank with a constant head for the period of the calibration. The orifice was constructed very carefully in accordance with fluid mechanics standards, and the flow through the orifice calculated from a knowledge of the static head and the orifice coefficient.

A typical trace from a hot-wire probe is shown in Figure 9.27. This is *not* an actual recording from this experiment. Figure 9.28 shows the results at three different speeds as the probe was moved up and down from the horizontal plane through the impeller center.

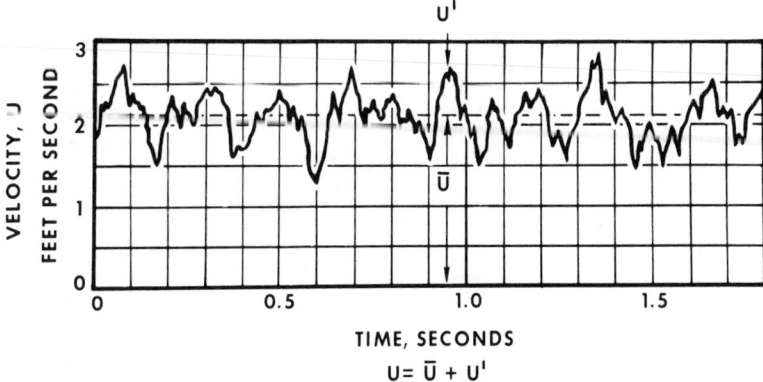

Figure 9.27 Typical velocity from hot-wire velocity probe. This is not a chart from this experimental program.

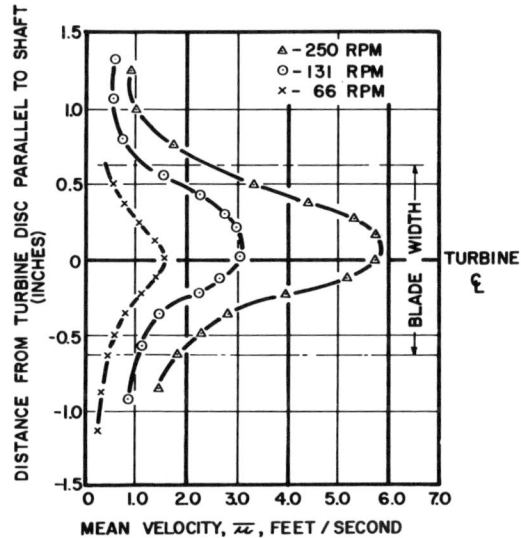

Figure 9.28 Mean velocity, $\bar{u}$, at various positions above and below a 6-in.-diameter turbine, six flat blades.

The average value of the slope of the mean velocity, $\bar{u}$, versus distance line is plotted in Figure 9.29. The maximum slope measured is also plotted in Figure 9.29. With the 6-in. flat-blade turbine, the maximum shear rate was approximately twice the average shear rate. However, as the impeller size was decreased or increased, the

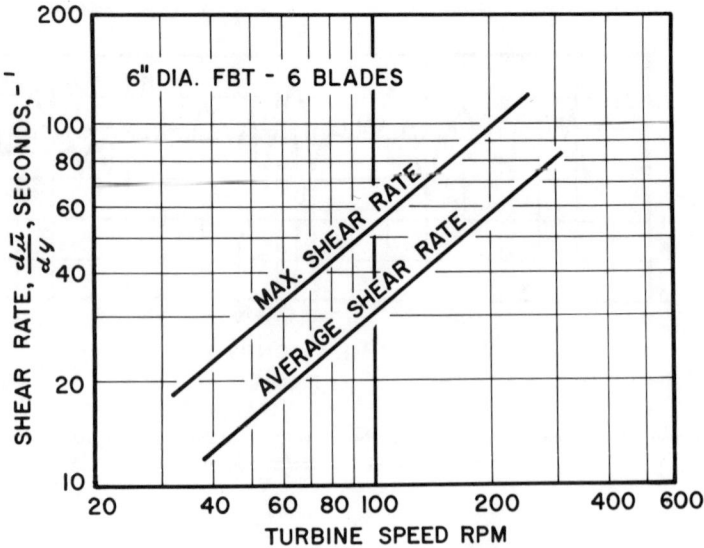

Figure 9.29 Maximum and average shear rates obtained from Figure 9.6.

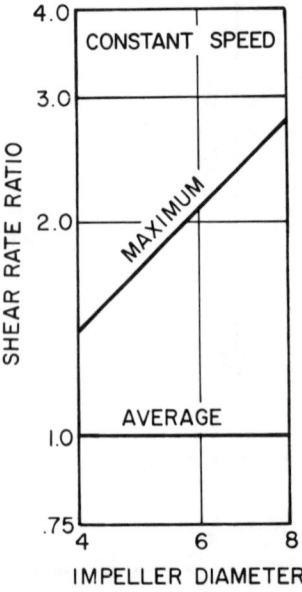

Figure 9.30 Ratio of maximum and average shear rates for 4-in., 6-in., and 8-in. flat-blade turbine impellers, compared to average shear rate for 6-in.-diameter impeller.

average shear rate remained about the same at the same impeller speed, Figure 9.30. The maximum shear rate increased with impeller diameter at the same impeller speed. This indicates that, as scaleup progresses, lower average shear rates will be encountered in the stream from the impeller due to lower impeller speeds, but higher maximum shear rates, due to the higher peripheral velocity of the impeller, Figure 9.31.

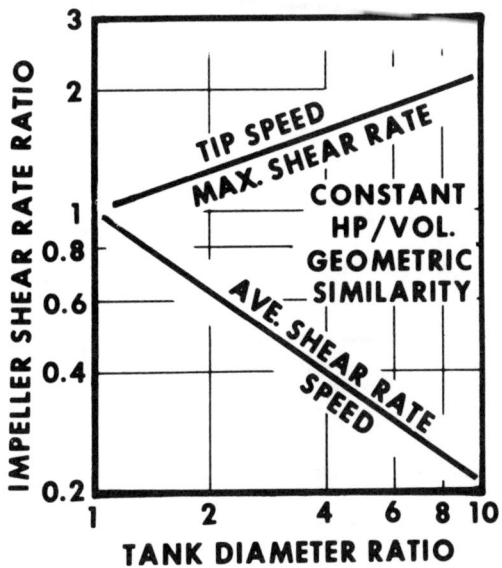

Figure 9.31 Typical changes on scale-up. The power per unit volume is held constant and geometric similarity is maintained.

The next question, then, is "What kinds of velocity fluctuations are found in this stream?" By taking the root mean square of the velocity fluctuations, one obtains a measure of the intensity of the turbulent fluctuations. This particular probe (and electronic equipment) considers all fluctuations having a frequency of 100 cps or higher.

To obtain a complete picture of the turbulent fluctuations and energy decay and dissipation, it would be necessary to actually analyze the complete spectra of frequency and size of the turbulent currents.

All the horsepower applied to the tank appears as heat, and since this is generated by viscous shear, when the eddy becomes small enough that its energy is dissipated by that mechanism. Therefore, processes that occur at the molecular level may well be a function only of energy input per unit volume.

As can be seen from Figure 9.32 the intensity of turbulence follows a pattern similar to the mean velocity. The ratio of the root mean square of the fluctuations to the mean velocity is between 0.4 and 0.6 within the jet from the impeller, Figure 9.33.

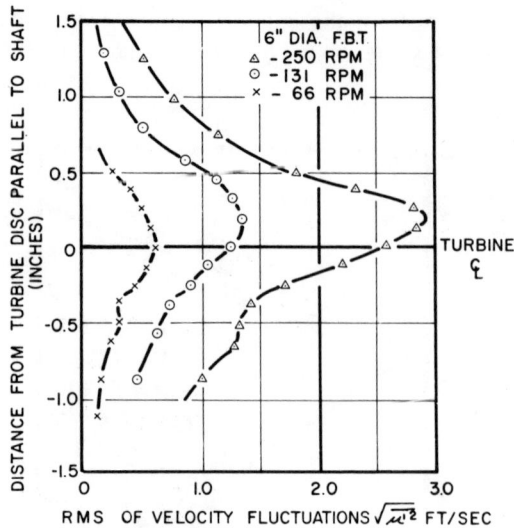

Figure 9.32 RMS of velocity fluctuations for 6-in.-diameter flat-blade turbine as shown in Figure

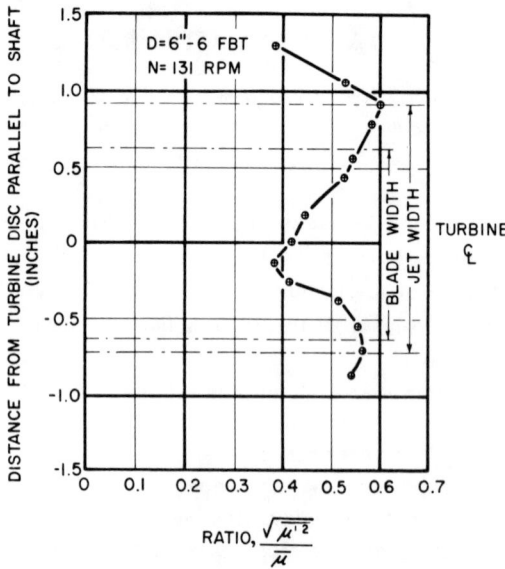

Figure 9.33 Ratio of RMS velocity fluctuations to mean velocity at corresponding points.

NOMENCLATURE

B	baffle width
b	maximum input concentration over average input concentration
C	off-bottom distance of impeller
c	concentration
$\bar{c}$	average concentration
c_p	specific heat
D	tank diameter
D_b	blade length
D_d	disc diameter
D_w	blade width
D_w/D	blade width-to-impeller dia. ratio
D/T	impeller diameter-to-tank diameter ratio
d	tube diameter
F	superficial gas velocity
FBT	flat blade turbine
g	gracitational constant
HP	horsepower
HPG	horsepower, gassed
h	heat transfer coefficient
h_o	heat transfer coefficient, no heating or cooling
k	thermal co ductivity
L	tank length
N	impeller speed
n	number of tanks
N_{Re}	Reynolds number, ratio of inertia force to viscosity force
P	power
P/V	power per unit volume of liquid
Q	flow from impeller per unit time
Q/V	flow per unit time per unit volume of tank
RMS	root mean square u'^2
RPM	revolutions per minute
T	tank diameter
t	designated time
u	velocity
u'	fluctuating velocity
$\bar{u}$	mean velocity over a time interval
V	tank volume
W	tank width
Y	height above impeller centerline horizontal plane
Z	liquid depth
Z/T	liquid depth to tank dia. ratio

μ viscosity
λ fluctuation period
ρ density
Θ residence time

REFERENCES

1. Sachs, J. P., and J. H. Rushton. "Discharge Flow from Turbine-Type Mixing Impellers," *Chem. Eng. Prog.* 50 (12) (1954).
2. Coyle, C. K., "Heat Treansfer to Jackets with Close Clearance Impellers in Viscous Materials," *Can. J. Chem. Eng.* (June 1970).
3. Oldshue, J. Y., and A. T. Gretton. "Helical Coil Heat Transfer in Mixing Vessels," *Chem. Eng. Prog.* 50 (12): 615–21 (1954).
4. Ibid, and H. E. Hirschland. "Blending of Low Viscosity Liquids with Side-Entering Mixers," *Chem. Eng. Prog.* 52 (11) (1956).
5. Strek, F., and S. Masiuk. "Heat Transfer in Liquid Mixers," *Int. Chem. Eng.* 7 (4) (1967).
6. Hubbard, P. "Operating Manual for II HR Hot-Wire and Hot-Film Anemometer," Iowa State University (1957).

CHAPTER 10

MIXING AND KNEADING EQUIPMENT – SOLIDS AND PASTES

RONALD W. REID
Charles Ross & Son Company
Hauppauge, NY

INTRODUCTION

Mixing and kneading equipment for solids and pastes mixing applications is available from a large number of domestic and foreign sources. The basic types of equipment used for mixing have been available for many years. Design variations are, however, introduced frequently to improve efficiency and meet changing market conditions.

Modern Industry considers mixing to be one of the most important steps in the manufacturing cycle. The success of this manufacturing operation plays a vital role in the overall success or failure of later operations. Even though this is such an important step in the manufacturing cycle, it is interesting to note that the engineering application of mixing equipment is more art than science, even in today's advanced technological society.

The mixing process usually involves two or more of the fundamental mixing operations listed below:

1. blending – mixing of two or more liquids or solids to homogeniety;
2. dispersion – mixing of two or more immisible components;
3. dissolving – to cause a solid to pass into a liquid;
4. disintegration – size reduction of solids;
5. emulsification – dispersion of liquid into a liquid;
6. heat transfer – mixing to control temperature;
7. mass transfer – mixing to react one product with another to form an end product; and
8. granulation – mixing of a solid and liquid to form an agglomerate.

Selection of equipment available to be used in a given process requires that one first evaluate which of the above operations are included in his process. Once this evaluation has been completed, an initial selection of commercially acceptable equipment can be made. Testing programs with manufacturers at their technical service facilities help to determine which of the prospective manufacturers' equipment is best suited to provide the desired process result.

In addition to evaluating which equipment is suitable to achieve a desired result, one must also consider several important selection criteria:

1. Batch vs continuous processing

2. Material flow and storage
3. Manpower requirements
4. Ancilliary equipment
5. Horsepower requirements
6. Downstream product use and desired form
7. Floorspace requirements
8. Cleaning requirements
9. Materials of construction and special design features to suit industry standards.

Included for discussion in this chapter are the major types of batch and continuous equipment used for solids blending and paste mixing.

SOLIDS BLENDING EQUIPMENT

Ribbon Blenders

Ribbon blenders (Figure 10.1) are primarily used for solids-solids blending; however, they may also be used to add liquids to solids or to blend fluids of medium-to-high viscosity. A dual helical ribbon is attached to a horizontal shaft, which is enclosed within a U-shaped trough or cylinder. The ribbon revolves slowly (15–60 rpm), depending on the size of the blender. Peripheral speeds of the agitator are approximately the same for all sizes and fall in the 250–350 fpm range. The ribbon moves materials in three directions:

1. the outer ribbon from the end of the trough to the center;
2. the inner ribbon from the center of the trough to the end; and
3. radially.

Discharge normally takes place in the bottom center of the trough. Continuous designs are available with the inner and outer ribbon moving materials the entire length of the trough and discharging at the end of the trough. A thorough interchange of material takes place in the trough within a relatively short cycle time. If solids of similar bulk density and volume are being blended, it is not uncommon to complete blending in 10 minutes or less. More difficult blends involving liquid introduction may take longer depending on the physical characteristics of the product being blended.

Ribbon blenders may be jacketed for heating or cooling and designed for internal pressure and vacuum operation. These features make the unit suitable for drying operations. Horsepower consumption and cost are relatively low per volume of material. Horsepower demand is based on bulk density, percentage of moisture, specific gravity and the speed with which the unit will be operated. Most manufacturers size their standard machine drives to handle material having a bulk density of approximately 35 lb/ft^3.

Units are available to over 500 ft^3 capacity. The minimum quantity of material that can be blended efficiently in a Ribbon blender is approximately 40% of the full-rated working capacity, because both the inner and outer ribbon must make contact with all materials. Since a wide range of sizes is available and scaleup

Figure 10.1 Ribbon blender — 10 ft^3 working capacity, 5-hp drive with shaft mounted torque arm reducer. Manufactured by Charles Ross & Son Company.

accurate, it is convenient to use the blender for both developmental and production requirements. Units may be mounted through the floor or on their own legs. Load cells to weigh charges, fixed material handling connections and dust collection equipment are easily attached. Bagging equipment, conveying stations or surge hoppers may be easily connected to large discharge openings in the bottom center of the trough.

Thorough cleaning of these machines is relatively easy if accessibility to the ribbon agitator is provided during installation. Packing glands of split design for complete disassembly and cleaning are available for sanitary requirements. Standard materials of construction are carbon and stainless steel. Special alloys and coatings may be provided when required.

Double Cone Blenders

Double Cone blenders (Figure 10.2) are generally used to blend free flowing solids. The solids may vary in bulk density and in percentage of the total mixture. The mixing action is very gentle; thus, easily degradable or heat-sensitive materials may be blended with minimal concern.

Construction is quite simple. Two cones are attached to a short cylindrical

Figure 10.2 Paul O. Abbe Interchangeable Body Rota-Cone Blender.

section. Support trunnions and drive shafts are attached to the cylindrical section. The shafts are, in turn, held secure by antifriction roller bearings that permit the entire double cone assembly to be rotated end over end until a homogeneous blend is achieved. In smaller sizes the container rotates at approximately 30 rpm and is turned by fractional horsepower motors. The larger blenders rotate at approximately 10 rpm and are driven by 25-hp motors. Larger drives are used as the density of the material increases. Horsepower consumption is relatively low when comparing these to other blenders that are designed to accomplish a similar result.

Standard models are constructed through 500-ft^3 full capacity. Working volumes are calculated at only 50–65% of the full capacity; therefore, a 500-ft^3 blender has an effective working volume of 325 ft^3. Optional features such as vacuum construction and jacketing for heating and cooling are available. Double Cone blenders are well suited for use in drying applications. Spray nozzle adapters may be added for the uniform addition of liquid ingredients to the main body of solids. High-speed internal agitation elements may be provided to disperse minor dry or liquid additives and to prevent or eliminate the formation of agglomerates.

Since the interior walls are open and unencumbered with rotating shafts or stuffing boxes, these machines are easily cleaned. Charging and discharging connections must be disconnected and reconnected between all batches. Nearly 100% of all materials in the blender will easily discharge on completion of the blend cycle. Design modifications are available that permit use of the cone after blending as a material storage device. In one case the cone may be split with the section containing the blended material used as a material transfer pan. In the second case the entire cone may be removed from a driver assembly and used as a hold bin. In this case the drive assembly may be used nearly continuously with multiple cones.

Twin Shell

The Twin-Shell V blender (Figure 10.3 was introduced in 1950 and has proved to be one of the most efficient batch blenders for providing uniform solid-solid or liquid-solid blends in critical applications. These include the blending of materials to make a wide variety of raw chemicals, pharmaceuticals, insecticides, cosmetics, plastics, etc.

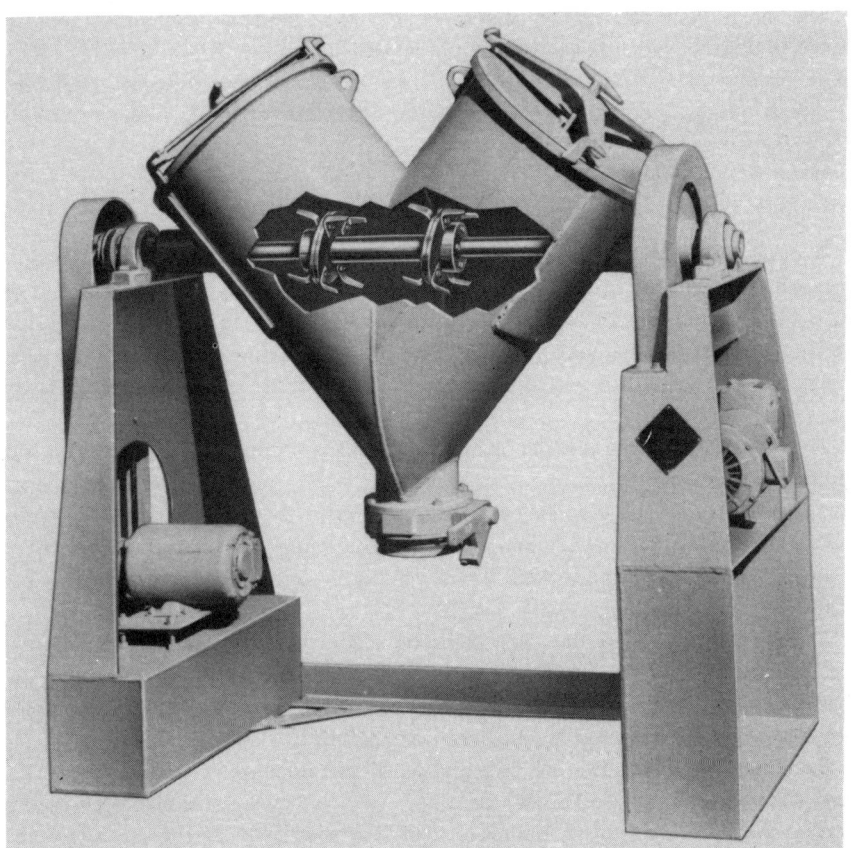

Figure 10.3 Patterson-Kelley Twin-Shell Liquid/Solids Blender.

The key to the Twin-Shell's blending efficiency is its design, which produces divergent material flow, causing the material to intermesh as it splits and refolds in the legs of the blender as it rotates.

The standard Twin-Shell model — without an intensifier bar — is used to gently blend dry, lump-free materials. However, intensifier bars are sometimes used to extend the capability of the Twin-Shell. For example, in dry blending, an intensifer bar can be used to break soft, friable agglomerates that are found in some raw materials or are formed during blending. Another bar, the high-speed intensifier bar, creates considerably more intensive action and is used to break down smaller, harder agglomerates, and it has a definite effect in shortening blending time.

The most versatile of the Twin-Shell blenders is the liquid-solids model. This unit uniformly disperses liquids into solids, resulting in use for a broad range of applications. It is widely used for dedusting solids, coating solids, agglomerating, controlling chemical reactions and blending all types of materials, from fine to coarse.

The Twin-Shell can handle the most difficult liquid-solids blending jobs and often eliminates several stages of mixing, pulverizing and screening to reduce equipment investment and materials-handling costs. Its consistent, uniform blends meet the most critical quality control standards. Standard capacities range from 1–75 ft^3. Larger units can be offered on request.

Cross-Flow Blender

The new Cross-Flow blender (Figure 10.4), introduced by Patterson-Kelley Company in late 1977, retains the basic V design of the Twin-Shell blender, but one leg of the blender is shorter than the other. This uneven leg design, which is based on the principle of unequal displacement, produces a unique and rapid axial exchange of material from each leg of the blender to the other. The result is very short blending times.

When the Cross-Flow blender is in the upright position, each leg holds an equal volume of material. But in one revolution, 25% of the material has been exchanged from leg to leg. As the blender rotates continuously there is a constant cross-flow shift of material. This also combines with the radial mixing action as the two inclined cylinders intermesh their flows. This divergent-flow concept was the basis of the Twin-Shell design.

The Cross-Flow offers the same blending accuracy, control and predictability as the Twin-Shell, but has the added advantage of improved discharge because of a sharper angle of the legs. Its design also allows for quick charging and discharging, easy access for cleaning and low horsepower operation.

P-K Cross-Flow blenders are constructed of carbon steel or stainless steel. Standard sizes range from 1–150 ft^3 capacity, with larger sizes available on request. Laboratory models in either stainless steel or transparent acrylic are available in 8-qt and 16-qt capacities.

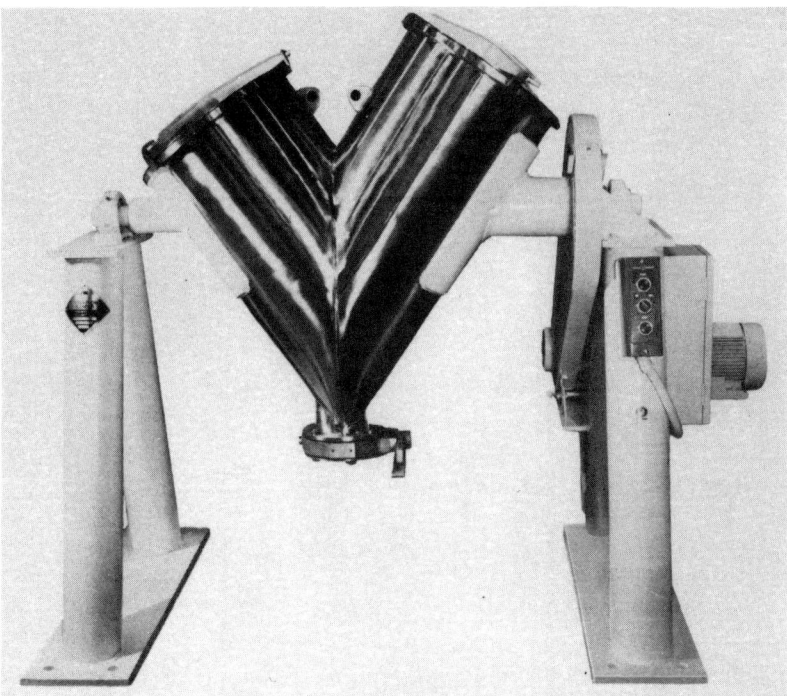

Figure 10.4 Patterson-Kelley Cross-Flow Blender.

Conical Screw Mixers

Conical Screw mixers (Figure 10.5) are efficient solids-solids blenders. They may also be used to blend solids-liquids and pastes of medium consistency. This mixer includes a unique screw flight agitator that revolves on its own axis and, at the same time, orbits the periphery of the cone. The screw rotates at approximately 64 rpm and orbits at approximately 3 rpm. Speeds may be increased or decreased slightly to compensate for the individual characteristics of the material being mixed.

During dry blending the screw operates in an up or forward mode. It continually brings new materials to the surface and distributes them in a spray-like pattern over the top of the batch. When blending pastes the opposite procedure is used, the screw is run in reverse pushing liquids and solids together with a combination of hydraulic and mechanical shear forces.

Conical Screw mixers are available in either single- or double-cone designs. Single-cone models are available through 353 ft^3 and double-cone models through 779 ft^3. Double-cone designs have two independent drive systems and mixing screws. Single-cone machines may be supplied with two screws where high capacity or especially critical mixing is required. Mixers of this design use low horsepower

Figure 10.5 Day Mixing Conical Screw Nauta Mixer.

drives. Since the machine is fixed mounted, it is convenient to attach material handling devices for both charging and discharging of materials.

Blending cycles are short and remain relatively constant as the working capacity increases. The screw diameter increases as does the capacity of the machine. Conical Screw mixers permit the efficient blending of batches as small as 10% of their rated working capacities. One machine can, therefore, potentially be utilized for wide-ranging production requirements.

Minimal floor space is required; however, headroom requirements in larger sizes can be substantial. The design lends itself to multistory or mezzanine installation.

Discharge is nearly 100% when used for dry, free-flowing materials. Complete cleaning can be difficult as access to all interior surfaces is limited.

High-speed lumpbreakers, jackets for heating or cooling, vacuum construction, spray nozzles, temperature probes and other options are available to make vertical cone mixers suitable for a wide range of requirements.

Hi-Intensity Mixers

Hi-Intensity mixers (Figure 10.6) are utilized extensively where rapid solids-solids or solids-liquid blends are required. Hi-Intensity mixers are available in both vertical and horizontal designs. They are both known for extremely short blend cycles and may be used for similar application requirements. The vertical model is available in selected sizes through 39 ft^3 working capacity. It employs a high-speed impeller with impact velocities of up to 150 ft/sec.

Figure 10.6 Welex Vertical Hi-Intensity Mixer.

The impeller forces circulation within the mixing container, which typically has inward sloping sides to minimize material hang-up. The impeller drives the materials upward and outward and also down in an intense vortex. The down draft of the impeller creates such an intense circulation that cavitation on dough-like materials is avoided. The combination of high-velocity agitation impact combined with interparticle friction ensures breakdown of agglomerates and uniform distribution of minute quantities of additives.

High turnover is achieved with these units. A 5-ft^3 mixer will typically be rated at between 900–1500 lb/hr whereas a 39-ft^3 mixer is rated at 9–11,000 lb/hr for a PVC dry blend. These machines are powered by 30- and 400-hp motors respectively. You can see that the power input per quantity of material is very high in comparison to other types of dry blending systems.

Vertical Hi-Intensity mixers may be jacketed for heating or cooling and may also be designed for vacuum operation. Units are constructed of polished stainless steel, thus cleaning is readily accomplished. Discharge is accomplished through a port mounted in the bottom side of the mixer.

Zig-Zag Continuous Blender

The P-K Zig-Zag blender (Figure 10.7), an out-growth of the Twin-Shell design, offers the blending accuracy of the Twin-Shell but on a continuous basis. Working with a gentle tumbling action, the Zig-Zag blends solid-solids or liquid-solids materials. Particles move by gravity, not by paddles, screws or baffles. The Zig-Zag repeatedly achieves intimate blends of 1 part in 5,000.

Figure 10.7 Patterson-Kelley Continuous Zig-Zag Blender.

The Zig-Zag is used for a wide range of applications, including simple multiple-component blends; complex catalytic coatings to produce reactions in chemicals, pharmaceuticals, ores, and plastics; controlling agglomeration, via adding a liquid binder to improve flow properties, stability, solubility of fine solids and reclaim fines; and numerous others.

Uniform blends are obtained through multiple recycling within the legs of the blender. At each half-turn, part of the material moves forward, part of it backward. Random splitting, merging, tumbling and rolling bring particles into contact with each other.

Comparatively short-term feed variations are leveled off by the time blended material reaches discharge, with recycling serving as an averaging device. With each revolution, the blender discharges a uniform quantity of material of constant volume and weight. Slope adjustment controls the material's residence time (average blending time), with most blends accomplished in three minutes or less. Cleaning is quick and efficient.

In liquid-solids blending, liquid injection into the tumbling, suspended mass of solid particles is achieved with the highly developed P-K liquid-dispersion bar. Metered liquid is discharged centrifugally as a mist of controlled droplet size. It impinges on particles, never reaching vessel walls. Processing conditions can be adjusted so that liquid addition imparts desired characteristics to the solid feed.

Time required to achieve a good blend is the same regardless of the unit's size. This has been verified through numerous tests and in all units in field applications. P-K Zig-Zag blenders are available in capacities from 10–10,000 ft^3/hr.

Continuous Blendex Motionless Mixers

The newest continuous dry solids blending technology of motionless mixing was commercially introduced to the marketplace in 1968. Motionless Mixers (Figure 10.8) are no moving parts devices" that split and recombine flow streams to provide perfect repeatable blends.

Blendex Motionless mixers are mounted vertically and are fed by means of gravimetric or volumetric dry solids feeders. Solids fed to this unit are passed through by gravity and discharged directly to downstream operations. The ideal solids being blended will have similar bulk densities and flow characteristics; however, may have widely different flowrates. It is possible to blend a major carrier with as little as ½% of a minor addition and still achieve a perfect blend. These units are ideal for the gentle blending of solids that are soft and friable and where it is important to minimize particle degradation. Since there are no moving parts, mechanical particle size reduction is completely eliminated.

Units are supplied in module design. Each module typically contains four or six mixing elements. A four-element module being charged with two raw materials will layer the two materials 512 times; a six-element module will layer the discharge material 8,192 times. The degree of mixing is predictable, thus easily controlled or varied using different element combinations.

Chapter 10. Ronald W. Reid

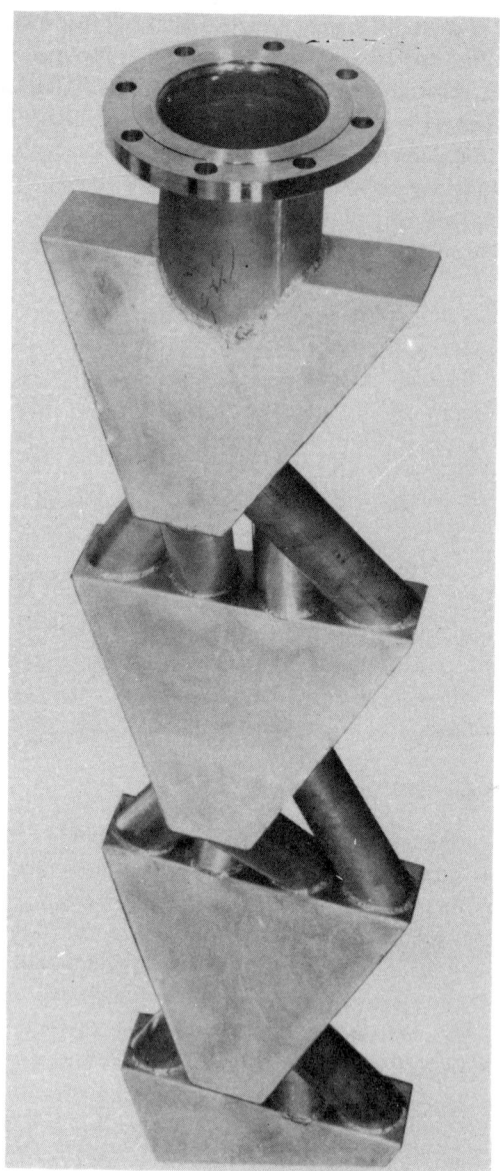

Figure 10.8 Ross Blendex Continuous Dry Solids Motionless Mixer.

Blendex dry solids Motionless mixers are available in either carbon steel or stainless steel construction and may be supplied through 8 in. diameter. The length of the various modules will vary depending on the diameter of the interconnecting tubes. For example, a 2 in.-diameter 4-element model has an overall length of 56 in. and is designed to handle approximately 6,400 lb/hr. A 4 in.-diameter 4-element unit has an overall length of 112 in. and is designed to blend approximately 26,000 lb/hr. Where cleaning is required it is possible to add removable cleanout ports between each mixing element.

Mullers

Mullers (Figure 10.9) are used to intimately blend minor amounts of liquid into a major solid carrier. A circular pan with a flat bottom contains a vertical shaft in the bottom center of the pan. Attached to the vertical shaft are horizontal extensions that hold free turning wheels which, in turn, rotate at slow speed around the periphery of the pan. Plows are attached to the shafts and direct the flow of material to the wheels, which then pass over the materials in the pan. The wheels do not touch the pan during the mixing cycle; their clearance is controlled for each application. As solids and liquids pass under the wheels, a smearing action takes place, much like that of a spatula working against the sides of a small container. An intimate dispersion of solids and liquids takes place. This unit is ideally suited for those applications where the generation of agglomerates is deemed totally undesirable.

Discharge takes place through a bottom or side door. Units are manufactured in a limited size range and are rated on cubic foot of discharged product. Specific gravity and the plasticity of the material to be mixed are the most critical factors in determining horsepower for a given application. Small units (5–7 ft^3) capacity require approximately 7½ hp whereas 60-ft^3 machines require up to 60-hp motors.

PASTE MIXING EQUIPMENT

Double Planetary Mixers

Double Planetary mixers (Figures 10.10 and 10.11) are primarily used to mix high-viscosity pastes. They are also used in solids-solids mixing applications as well. The unit is of change-can design (the stirrers may be raised out of the mix vessel by means of a self contained hydraulic lift). During the mix cycle, two rectangularly shaped stirrers revolve around the mix tank on a central axis. Simultaneously, each blade revolves on its own axis at approximately twice the speed of the central rotation. With each revolution on its own axis, each blade advances forward along the tank wall. Complete homogeneity of all materials being mixed is achieved within a few short minutes. The short cycles can be attributed to the positive movement of the stirrers on all materials contained within the mix can.

During paste blending a let-down procedure is typically used. All solids are charged to the tank and premixed. On completion of premixing, approximately 1/2–2/3 of the total liquid available for the formula is added to the batch. The machine is then run until a highly viscous, dough-like material is produced. After a

Figure 10.9   National Engineering Company Simpson Mix Muller Model 1½F

short time, the batch is then let down to its final consistency with the remaining liquid. Complete cycles can usually be achieved within 20–30 minutes.

Double Planetaries may be used to mix approximately 90% of the viscous solid-liquid products being manufactured today. For those applications that require high shear to masticate or reduce large plastic or rubber solids, other equipment should be considered. Batches as small as 10% of the rated machine-working capacity may be worked efficiently in this mixer.

Standard sizes range from small one-gallon laboratory units to 300-gallon production machines. There are no submerged seals or stuffing boxes in the product zones, thus cleaning or color changes may be readily accomplished.

Multiple mix cans permit mixing in one can while charging and preparation for a second batch is taking place in another. Nearly 100% utilization of this machine can be realized in this manner.

Optional features such as jacketed mix cans for heating or cooling, scrapers to continuously scrape the tank walls and vacuum construction are available. These mixers are known for their low horsepower consumption per volume of material mixed.

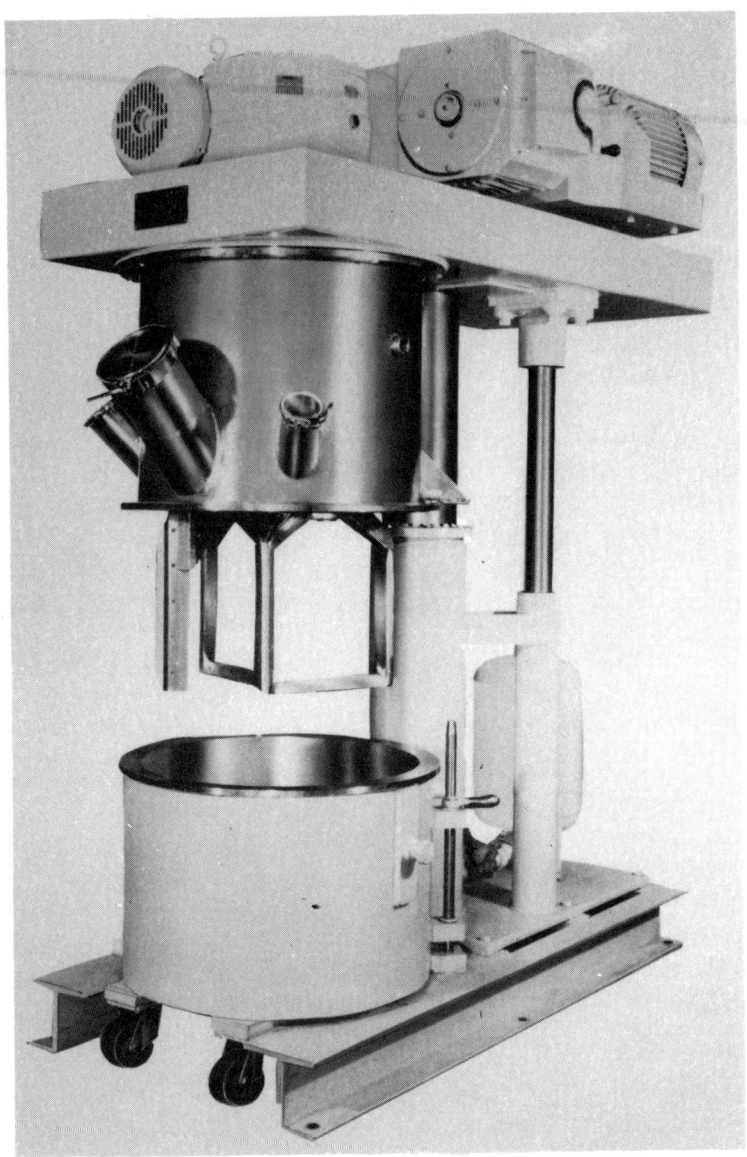

Figure 10.10 Ross HDM 25-gallon Double Planetary Mixer, vacuum, jacketed, variable speed drive.

Chapter 10. Ronald W. Reid

Figure 10.11 Mixing Pattern Ross Double Planetary Mixer.

Double Arm Mixers

Double Arm mixers (Figure 10.12) are rugged, heavy duty machines for high-viscosity mixing requirements. These mixers contain two mixing blades that operate opposite each other in a U-shaped horizontal trough. The blades rotate toward each other at differential speeds (approximately a 2:3 ratio). Blades may be mounted for either tangential or overlapping operation. Tangential designs are used when the feed materials are in sheets or chunks and tend to lie on top of the agitators rather than flow down between the blades. Overlapping blades are used on materials that will flow between the agitator blades. Numerous blade designs are available; however, the vast majority of mixers incorporate either sigma or spiral blade configurations.

Double Arm mixers have the strength and power to tear apart and reduce solid particles from small pellet-sized resinous materials to large blocks of rubber. An intensive shearing action takes place between the close tolerance blades and trough wall. Ram covers are sometimes used to increase the shear capabilities and thus reduce mixing time. Double Arm mixers are rated on a gallon-working capacity and full holding capacity basis. For efficient mixing to take place, it is necessary for a machine to have a minimum charge of approximately 40% of its rated working

Figure 10.12. Double Arm mixers. Day Mixing 200-gallon non-tilting design.

capacity; therefore, a machine must be carefully sized for a given application.

Standard sizes are available through 1,000 gallons. Jackets for heating or cooling, vacuum construction and numerous other options may be supplied to extend the versatility of the mixer. The discharge of conventional Double Arm mixers is usually accomplished by tilting the trough 90° from its working position and dumping its contents into tubs or onto the floor. Material handling can be difficult with materials that have little flow or elevated temperatures at discharge.

A design variation is available that improves the mixing efficiency and discharge capabilities of Double Arm mixers. The Kneader Extruder (Figures 10.13 and 10.14) introduced in 1958, not only includes the heavy-duty mixing blades, but also an extrusion screw in the cavity beneath the mixing blades. The extrusion screw rotates during both the mixing and discharge cycles. During the mix cycle, the screw continuously moves materials within reach of the mixing blades. This action helps to shorten mix cycles by up to 20% compared to conventional Double Arm mixers. In addition, the screw assists in cutting and reducing solids. Once the mixing cycle is completed, the discharge screw is reversed and utilized for emptying the machine. Average discharge times vary from 4—12 minutes depending on die configuration.

Banbury Mixer

The Banbury mixer (Figure 10.15) is a batch-type mixer capable of mixing

Chapter 10. Ronald W. Reid

Figure 10.13 AMK Kneader Extruder, screw discharge design.

thick, viscous and hard-to-mix materials in a very short period of time. While the Banbury has performed such diverse tasks as mixing pizza dough and grinding pine tree stumps, it is most commonly used for the breakdown or incorporation of pigments and fillers into rubber and plastic materials.

The heart of each Banbury is the mixing chamber, which contains two rotors of unique design. Materials to be mixed are fed into the Banbury at the top. The rotors are driven by a high horsepower electric motor while pressure is applied from the top by a plunger or ram cover. In a very few minutes the materials are thoroughly mixed and are discharged through a door at the bottom.

A Banbury is said to mix in four ways: milling, kneading, longitudinal cut-back and lateral overlap. Milling is described as the action of the rotors wiping material through the periphery of the mixing chamber. The mass in front of the rotor blades is subjected to rolling action, which causes heavy molecular shear in the layers of mass. Kneading is caused by the rotor tips when material that has been deformed by milling is carried to the center of the chamber and relaxed. Longitudinal cutback is accomplished by the rotor blades, which are designed to continually force material to the center of the mixing chamber. Since each rotor is designed with two

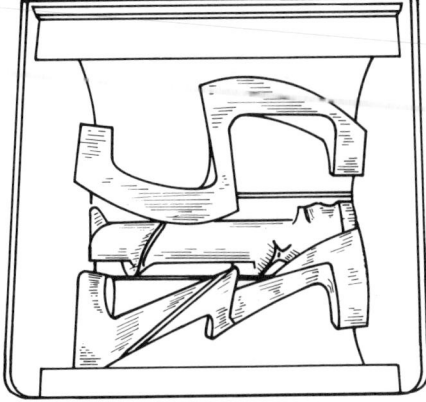

Figure 1 — Overhead view of Kneader-Extruder showing sigma blades and extrusion screw.

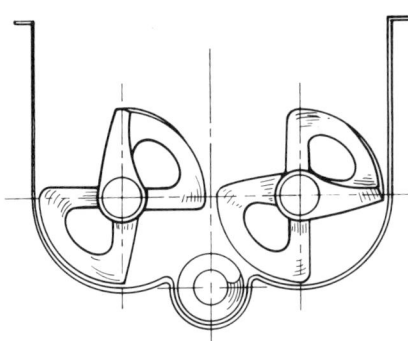

Figure 2 — Side view of Kneader-Extruder showing tangential relationship of sigma blades and extrusion screw.

Figure 10.14. Illustrations of Kneader Extruder interior.

opposing spiral tips of unequal lengths, a longitudinal smearing and rolling action takes place. Lateral overlap is the action by which a portion of the material in front of one rotor is pushed over into the opposite chamber or side and then is worked in with material in front of that rotor until some of the material is pushed back to the other rotor.

Typical clearances between the rotor tip and chamber wall are 3/32—3/8 in., depending on the size of the mixer. Banburys are designed with extremely high horsepower motors to accomplish a given job in the shortest possible cycle time. It is important, therefore, that an operator control closely those factors that play an important part in the desired end result: (1) mixing temperature; (2) rotor speed; (3) batch size; and (4) ram pressure.

Continuous Mixers

Continuous mixers (Figure 10.16) are used extenisvely for the manufacture of

This size F270 Uni-driven Banbury mixer has a gross capacity of 270 litres for typical tire batch of 500 pounds. It features larger feed opening, capsule drilled sides, hydraulic dust stops, two- or four-wing rotors and quick-acting drop door suitable for up to 2,000 hp at 60 rpm.

Figure 10.15 Banbury mixer size F270, 500-pound capacity.

high-viscosity pastes that are required in large volumes. Continuous mixing systems generally contain either single- or dual-rotor assemblies that are contained within long, cylindrical housings. These provide a combination of rotary and axial motions within the mix chamber, thus generating intimate, repeatable blends. Screw diameters vary with required output from small 2-in.-diameter pilot plant size through large 36-in.-diameter production size. Typical L/D ratios of 8:1 or 10:1 are found in this equipment. The rotors are designed to transfer materials from the feed hopper to the discharge end and to move all materials contained within this mix

Figure 10.16 Continuous mixer. Baker Perkins Ko-Kneader.

zone at a similar rate, thus providing a uniform retention time. This permits close control of shear, temperature and size reduction of the raw materials.

The cylindrical rotor housing may be provided with single- or multizone jackets that provide the user with a tool to accurately control temperature of the mixture. Chambers may be designed for vacuum operation and also for the addition of minor reactant materials. Drives of Continuous mixers are generally of multiple-speed design to allow the operator to coordinate the Continuous mixer production rate with upstream and downstream auxiliary machinery.

Continuous mixers are best suited for use with raw materials that are easily fed into the feed chamber. This requirement thus restricts the application of these machines to blends of liquids and solids that are of a nature to allow continuous, uniform feed into the feed hopper. Numerous dry solids and liquid weighing and feeding systems are available for the purpose of feeding these machines. It is important that sufficient time be spent in the selection of this auxiliary equipment as the success of a continuous system lies in the reliability of the equipment that is used in the feeding of raw materials.

Units are available in the usual range of standard and special alloys. Optional split-rotor housing designs are available providing for relatively quick access to the mix chamber for cleaning and maintenance purposes. Continuous mixers of this type require relatively high horsepower drives; however, they are not generally subject to the typical power surges one would expect in batch mixers. As a result, power demands can be minimized and savings in power costs realized.

Three Roll Mills

Three Roll Mills (Figures 10.17 and 10.18) are used to grind or disperse agglomerates of material. Three parallel, hardened steel rolls are mounted in a horizontal or inclined plane. The end rolls can be moved back and forth from the center roll by either manual or hydraulic adjustments. The manual method, such as the handwheel type, is generally accepted as the most accurate, although requiring somewhat more skill than the hydraulic type.

Figure 10.17 Ross/GPE 4 x 8-in. laboratory model Three Roll Mill.

The back roll is the slow-speed roll and revolves at approximately 40 rpm. The middle roll runs opposite to the back roll at three times its speed or 120 rpm. The front roll revolves three times faster than the middle roll, again in the opposite direction. Material flows from the back to the front roll where it is then removed by a razor knife on a metal plate called an apron. The apron funnels the material into a container.

Materials on the rolls are held in place by endplates that rest on the surfaces of the rolls. The surfaces of the rolls are finely ground and care must be taken not to damage them. As material is transferred from roll to roll, it becomes subjected to a number of shearing actions that take place at the nip of the rolls. As the rolls revolve in opposite directions to one another, the materials are broken down by a crushing action and the agglomerates of the unmixed materials are smeared into a film of material on the surfa e of the rolls. This material flows off the mill as a paste. Many materials require more than one pass.

With the intensive shear action that takes place, heat is often generated. This can be controlled by water cooling the center of the hollow rolls. Occasionally, materi-

Figure 10.18. Coxis 6 x 40" incline design Three Roll Mill.

als are heated on the rolls, this can be accomplished by hot water or steam. The rolls are generally made of carbon steel, chilled iron or stainless steel. A surface hardness of Rockwell C 58—60 is common.

Mills vary in size from 2½" X 5" to 16" X 46" production sizes. The small mills have an approximate output, depending on the gap setting, of 1—2 gal/hr whereas the larger mills can range from 100—200 gal/hr. The materials that are processed on the mills vary from a low viscosity of a few hundred cps to a high viscosity of a few hundred thousand cps. A wide variety of materials can be processed.

Vertical Tank Mixers

Vertical Tank mixers (Figure 10.19) of relatively simple designs have traditionally been utilized for the production of heavy pastes. Designs are available that bring together various combinations of agitation systems to optimize efficiency for a particular requirement.

Shear Bar Design — Rotating arms are attached to a rotating vertical shaft. The arms are positioned at an angle to provide upward and downward movement within the mix chamber. Rotating arms are offset from the one immediately below or above, thus reducing any shock loading on the drive components. In addition, there are fixed stationary arms welded to the vessel wall at close clearance to the rotating arms. This combination provides high shear and assists in breaking up lumps of solid materials.

SHEAR BAR MIXER

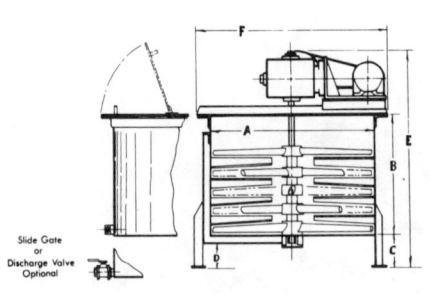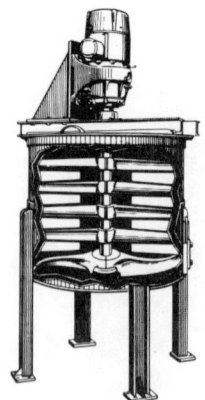

RAKE GATE MIXER

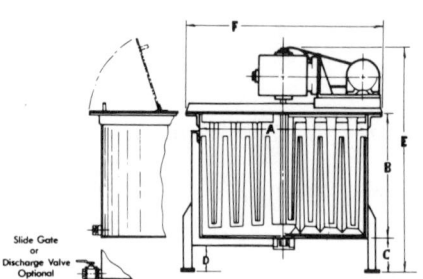

Figure 10.19

Rake Mixers — The rotating arms in this design are also attached to a vertical shaft. In this design, however, the arms are mounted vertically and mesh with stationary arms, which are affixed to the top of the vessel. This design is generally used for medium-viscosity mixing requirements. Rake mixers are well suited for use where scraping of the tank walls is required for heat transfer purposes.

Additional agitator designs are available that include both slow- and high-speed dispersion devices (Figure 10.20). Combination units permit a great deal of latitude in processing operations. A unit that includes both an anchor agitator and a high-speed dispersion blade may be used for initial high-speed/high-shear solids reduction, emulsification and also high-viscosity blending. Combination mixers maximize the efficiency of different mixing technologies and extend the useful range of each to provide short cycles.

Tank mixers may be supplied in the usual range of construction materials and are available to over 2,000-gallon capacity. Features include jackets for heating or

Figure 10.20 Ross VersaMix, 500-gallon, vacuum, internal-pressure, jacketed can.

cooling, vacuum or internal pressure designs, multiple-speed drives, special openings, finishes and discharge arrangements to suit each unique processing requirement.

PROCESS EQUIPMENT SERIES

Volume 1   Index

A

Absolute filter media	169
filtration	110
Accordion-type pad filter cartridge	165
Acrylic acid	49
Active filtration areas	77
aerator	247, 248
Aggregates	7
Air blowback	54
flotation	22
permeability	195
Alum	39, 48
Aluminum	8, 49
Anionic polymers	49
Asbestos fiber	115
ASME Code	188
Axial flow turbine	239, 248
pumping	240

B

Backwashing	1, 33, 36, 107, 130, 161
Baffles	240, 255
Bag filter	162
vessel	162
media	195
medium	191
Banbury mixer	283, 286
Basket centrifuge	89
Batch filtration	54
settling	4
Bleedthrough	172, 173
Blending process	244, 253
Blowback air	65
BOD	32, 38
Boiler	1
feedwater	1

Braid filter septa	130, 131
Bridging	48
Brownian movement	7

C

Cake discharge	61, 74, 136
forming time	52
resistance	51, 115
wash	62, 132, 143
Calcium	49
salts	49
Capital cost	112
Carbon bed filter	168
Cartridge	106, 146, 152
construction	140
filter	145
housing	151
Cellulose	115
filteraid	113
Centrifugal discharge/filter	131, 143
force	81
Centrifugation	44, 47
Centrifuge	81
cost	104
system	83
Ceramic cartridge	155
filter	123
Chemical coagulation	39
floc	2
plant	22
Chlorination	6
Clarification	2
Clarifier	222
Classification mechanism	176
Close-clearance impellers	246
Closed filter bag system	197
Cloth belt discharge	68, 70
filter septa	127
Coagulation	9, 39
Coil spring discharge	68
filter	68, 69
Colloid	7

Colloidal suspension	2
Column settling test	220
Compression rolls	57
zone	5
Conical basket	86, 87
screw mixer	273
Constant-stireed tank	10
Continuous filtration	52, 55
mixer	285, 287
vacuum filter	55
Critical solid loading	20
Cross-flow blender	272

D

Deep bed filter	166, 167
Depth filtration	110
Desludging centrifuge	94, 95
Detention time	10
Dewatering	30, 48
zone	42
Diaphragm positioning	138
Diatomaceous earth	36, 37, 115
Diatomite	113, 115, 116
filteraid	116
Diffusion	37
Discentrifuge	92, 95
filter	74
Discharge mechanism	67
Discrete settling	14, 15
Disposable filter media	107
Disposable dry cake	107
Domestic waters	1
Double arm mixer	282
cone blender	269
planetary mixer	279, 281
Drainge grid	59
member	51
Driving force	51
Drum filter	55
vacuum filter	71
Dry cake	107, 109
discharge	132, 133, 138

leaf discharge leaf filter	138
solids motionless mixer	278

E

Edge filter	164, 165, 166
EPA	1, 26
Equipment design	215
Evaporation	3

F

Fabric cloth	127
screen sleeves	157
Feed entry	217
Felt inspection	201
sealing	185
strainer bag	175
media	176
Fiber-wound filter tubes	147
Filter drum	56
leaves	138
media	36, 52
medium	78
pad	164
press	134
design	134
septa	112, 125
valve	57, 58
vat	60
Filteraid	109, 113, 116, 123
Filtering centrifuge	81
Filtrate outlet	52
Filtration	30, 35, 51
centrifuge	97
cycle	56
equation	51
First order kinetic	11
Fixed solids	3
Flat blade turbine	240
Flexible tube filter	130
Floc	7, 26, 36
Flocculant settling	14, 19

Flocculation	0, 35, 104
Flotation	21, 22
Flow distribution	216, 217
pattern	237, 259, 260
Fluid mixing	236
equipment	245
motion	257
properties	248
shear	243, 258
Fouling	1, 22, 26
Froth	24

G

Glass microfiber filter tube	154
Gooch crucible	3
Graduated cylinder test	220
Grass plots	40
Gravel filter	167
Gravity separation	2, 24
settling	14

H

Head loss	33, 36
Heat exchanger	1
transfer	251, 255
coefficient	256
Heavy metal	2
Heel filtration	143
Hi-intensity mixer	275
High viscosity application	243
Hindered settling	14, 20, 21, 221
Hopper dewaterer	73
Horizontal belt	55
endless cloth	75, 78, 79
leaf filter	139, 141, 142
plate filter	140
rotating pan	75, 77
tank filter	120, 121
tray filter	143
vacuum filter	75

Hydraulic loading	25, 33, 34
shock filter	128
Hydrophilic	7
Hydrophobic	7
colloids	8

I

Imhoff cone	4
Impeller	242, 248
flow	243
power consumption	250
Inclined plate settler	211
In-depth filter	39
Indexing cloth	79
Inerts	3
Inline filter	159
Inorganic salts	4
Internal drums	73
Iron	8, 48, 49

K

Kneader extruder	284, 285
Kneading	267
equipment	267

L

Laboratory tests	102
Lamella Gravity Settler	217, 226
Laminar flow	25, 28, 32
Leaf spacing	138
Lime	48
Liquid dispersion	252
liquid mixing	252, 253
solid blender	271
separation	81
pressure filter	105, 108
Lineblender	246
Low viscosity materials	253

M

Macrofloc	48
Mass transfer	253
Materials of construction	224, 226
Mechanical sluicing	119
Mechanized filter press	136
Membrane filter	169
Metal salts	48, 49
Micrometer rating	111
Microscreen	33
Microstrainer	32
Migration	172
Mining industry	22
Miscible liquid	253
Mixer speed	236
type	237
Mixing	6, 236, 267
design	237, 239
process	236
vessel	243, 258
Moore filter	54
Mother liquor	62
Motionless mixer	277
Moving bed filter	39
Mud discharge filter	170
Mullers	279
Multiple filter	160

N

Nauta mixer	274
Non-Newtonian fluid	254
Nonsettable solids	4
Nozzle centrifuge	94, 96
Nylon	40

O

Open filter bag system	200
Operating costs	112
Organic polymers	48
polyelectrolytes	49

O-ring design	186
Oscillating centrifuge	90
Overflow	16

P

Paper industry	22
Particle classification	175
Paste	267
mixing equipment	279
Peeler centrifuge	89
Perforate basket centrifuge	103
Perlite	115
Petrochemical plants	22
Petroleum	1
pH	6
Plastic woven cloth	127
Plasticized diaphragm	138
Plate and frame filter	40, 133
inclination	218
settler	211
design	215
theory	212
Pleated cartridge	149
paper cartridge	149, 150
Plug flow reactor	10, 11
Polyelectrolyte	8, 50
Polyester	40
Polymer	48
coagulant	39
Polypropylene	40
felt sleeve cartridge	158
Porous carbon	123
ceramid filter	124
media	123
tubes	123
Precipitate	2
Precoat filter	65, 115, 116, 132
Prefilter	169
Pressure drop	234
chart	235
filtration	40, 107
leaf filter	109, 117, 119, 135, 139
Primary clarification	6
Propeller unit	237, 238

Pumping capacity	242, 258
Push-type centrifuge	83, 84

R

Radial flow turbine	239
Rake mixer	290
Rapid mix	12
sand filter	39
Receiver	60
Reciprocating vacuum tray	79
Removable-medium filter	67
Residence time	27
Retention time	47
Reuseable media	111
Reverse osmosis	6
Reynolds number	251
Ribbon blender	268, 269
Riveted filter leaf	118
Roll discharge	65, 66
mill	288
Rota-Cone blender	270
Rotary disc	55
vacuum filter	73
drum	55
vacuum filter	63, 64
vacuum filtration	57
Rotating leaf pressure filter	121, 122

S

Saran	40
Scaling	22
Scavenger plate	140
Scraper discharge	61, 64
Screen bowl centrifuge	93
Screening centrifuge	88, 92
Scroll discharge	75, 76
screen centrifuge	86, 87
Sealing device	208
Sedimentation	2, 35
centrifuge	90, 99

Self-cleaning strainer	232
Selection parameter	206
Separation process	177
Series straining	233
Settling	1
Shallow depth sedimentation	213
Shear bar design	289
rate	242
Sheet filter	163, 164
Short-circuiting	24, 25, 26
Side-entering mixer	245
Sieving mechanism	110
Silica	49
Simpson Mix Muller	280
Single cell rotary drum filter	70
Sizing method	219
Slow mix	12, 19
sand filter	38
Sludge	1, 219
collector	22
compression	220
dewatering plant	135
withdrawal	219
Sluice headers	120
Slurry concentration	52
Snap-Ring®	179, 181
Sodium aluminate	8
Solar heating	26
Solid bowl centrifuge	91
Solids blending	268
Specific resistance	52
Stainless steel cartridge	154, 155
Stokes law	18
Strainer bags	175
concept	178
selection	228
Straining	35
String discharge	65
Surface aerator	249
loading	26, 20, 22, 25
theory	211
water treatment	227
Suspended solid	2, 3, 4

T

Tank baffles	241
mixer	290
Tee type strainer	229
Thermal properties	256
thickeners	222
Three roll mill	288, 289
Throwaway media	111
Tilting pan discharge	77
Top feed filter	72
Total settling area	214
solids	3
Traveling belt	78
Trickling filter	34
Trommel® belt	72
filter	71
Trunnions	270
Tube settlers	27, 28, 211
Tubular centrifuge	96, 97
element backwash filter	123, 159
filter element	123, 130, 144, 145
Turbine installation	258
Turbulence	24, 26, 31
Twin-Shell V blender	271
Two-receiver system	63

U

Ultrafiltration	43

V

Vacuum disc filter	74
filter	40, 42, 43, 51, 84
flotation	21
leaf filter	54
Nutsche	54
pump	52
receiver	52
tray	78
Van der Waals forces	7

Variable batch sizes	236
Vapor retaining hood	63
Velocity fluctuation	264
gradient	26
profile	243, 244
probe	261
Vertical pressure filter	126
tank leaf filter	137
mixer	289
Vibrating screen	32
Viscosity	243, 249
Volatile	3
Vortex clarifier	96, 98

W

Wash assembly	58
Wash blanket	57
liquor	62
Washing cycle	54
Waste disposal	1
treatment	1
Wedge wire elements	126
Weir	24
loading	27
Wet cake discharge filter	117, 121
Wire cloth	118, 156
Wound cartridge	147
wedge wire	125
Woven screen	127

Y

Y-Type strainer	229

Z

Zeta potential	7, 8
Zig-Zag continuous blender	276

RAYMOND H. FOGLER LIBRARY

DATE DUE